DISCOVERY of REALITY

DISCOVERY *of* REALITY

THE LIGHT OF SYSTEM PHILOSOPHY

SCIENCE AND PHILOSOPHY

GEORGE LUKE

Copyright © 2019 by George Luke.

ISBN: Softcover   978-1-5437-0452-5
       eBook   978-1-5437-0451-8

All rights reserved. No part of this book may be used or reproduced by any means, graphic, electronic, or mechanical, including photocopying, recording, taping or by any information storage retrieval system without the written permission of the author except in the case of brief quotations embodied in critical articles and reviews.

Because of the dynamic nature of the Internet, any web addresses or links contained in this book may have changed since publication and may no longer be valid. The views expressed in this work are solely those of the author and do not necessarily reflect the views of the publisher, and the publisher hereby disclaims any responsibility for them.

First published by PGL BOOKS
as eBook in Amazon Kindle Store in July 2018

Print information available on the last page.

To order additional copies of this book, contact
Partridge India
000 800 10062 62
orders.india@partridgepublishing.com

www.partridgepublishing.com/india

ALSO BY GEORGE LUKE

SAPTALOKADARSHANAM SAMGRAHAM (MALAYALAM)
JEEVANUM PARINAMAVUM – SYSTEM PHILOSOPHIYUDE VELICHAM
   (MALAYALAM)
ORIGIN OF UNIVERSE : THE LIGHT OF SYSTEM PHILOSOPHY
LIFE AND MIND : THE LIGHT OF SYSTEM PHILOSOPHY

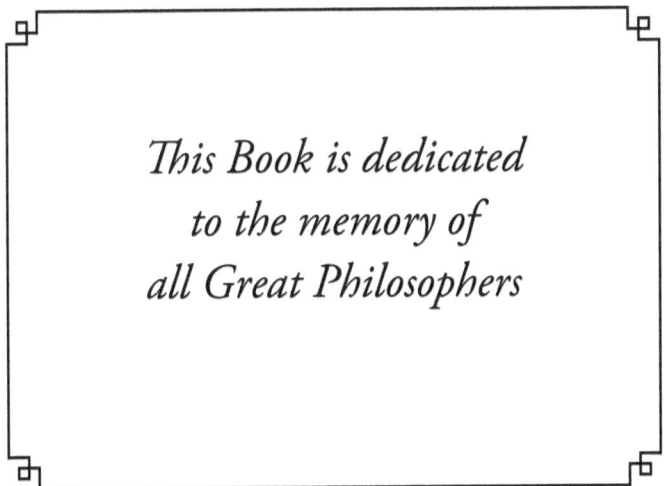

*This Book is dedicated
to the memory of
all Great Philosophers*

CONTENTS

Preface xi

Prologue xvii

Introduction xxv

Chapter 1: A Guide to the Levels of Knowledge 1

    1.1 Definition of Knowledge and its Classification 2

    1.2 The Organic Levels of Propositions 9

    1.3 Approach to Theory of Knowledge 12

Chapter 2: Functional Classification of Philosophy 14

    2.1 Objectives of Philosophy 16

    2.2 Ontology 20

    2.3 Theory of Knowledge (Epistemology) 23

Chapter 3: Worldviews and Versions of Reality 34

    3.1 Six Worldviews in Western Philosophy 35

    3.2 Organic Worldview (OWV) 38

    3.3 Spiritual Process Worldview (SPWV) 41

    3.4 Mechanistic Worldview – MWV 45

    3.5 Physical Process Worldview – PPWV 48

    3.6 Seven Theories of Reality 52

Chapter 4: Western Philosophy - Key Doctrines of Great Philosophers 58

4.1 Idealism of Plato 59
4.2 Idealist Philosophy of Aristotle 67
4.3 Spiritual Process Philosophy 70
4.4 Theory of Knowledge – Rational Mysticism 80
4.5 Theory of Knowledge – Empirical Mysticism 82
4.6 Deism and Rationalism 89
4.7 Materialism and Empiricism 97
4.8 A Note on Eastern Philosophy 104

Chapter 5: System Philosophy of Reality 109

5.1 Dilemmas about Reality 110
5.2 Three Levels of Natural Systems 113
5.3 System Model of Ultimate Reality 119

Chapter 6: Seven Life Systems and Knowledge 128

6.1 Social World and its Components 129
6.2 Existence of Seven Life Systems 133
6.3 System Model of Knowledge 142

Chapter 7: Comprehensive View about Truth 148

7.1 The Dilemmas about Scientific Truth 150
7.2 Issues about Religious Truth 157

   7.3 System Philosophy of Truth 157

Bibliography 171

Index of Names 179

Index of Subjects 183

Preface

It may be recalled that *science* in the strict sense originated only about five centuries ago. Scientific knowledge is produced in our particular mental faculty, which can be described as intellectual mind having the practical aims of life. This intellectual faculty is alternatively termed as *reason*. Science tries to understand the external things by observing the world in *physical* terms as revealed by the sense organs. Here, the word 'physical' refers to the measurable properties of matter and energy, which are inter-convertible. Further, scientists are in the pursuit of discovering theories of cosmology to explain the origin of physical universe. This enquiry boils down to the question about the primordial source of matter.

Deliberating upon the origin and evolution of natural universe, we cannot ignore the existence of nonphysical aspects - design, purpose and creativity. Traditionally, people believed that these aspects are due to the work of certain supernatural beings. The notion of God represents the supreme form of such supernatural beings. Then, in the course of time, the organized forms of religion emerged such as Christianity, Hinduism, Islam and Buddhism. The abstract interpretation about God and other notions of religious worship can be called *faith*. The characteristic feature of faith is the shade of mysticism that includes symbolism, imagination and metaphor.

The gulf between reason and faith – science and religion – appears to be very wide and unbridgeable. In this situation it is the chief aim of philosophy to synthesize these opposite kinds of knowledge. Philosophers of all times have tried to develop a vision about the original cause or **reality** of universe. The doctrines of this branch of philosophy are generally called *ontology* of which the hitherto course is controversial and full of logical inconsistencies.

This book adopts an innovative path for the philosophical enquiry into the longstanding question: what is the reality – the original cause or fundamental stuff – of universe? It adopts the intellectual faculty in contrast to the mystic view of religion. So the book introduces a new philosophy, named as System Philosophy, for synthesizing the diverse knowledge under science and religion. This is the reason for the title *Discovery of Reality*. And, the present book is the sequel to the previous two books *Origin of universe* and *Life and Mind*.

System Philosophy is to be distinguished from the *systems philosophy* or *systems view*, popularized in the writings of a group of thinkers including mainly David Bohm, Fritjof Capra and Ervin Laszlo. They have followed the empirical approach adopting the process view for articulating that the universe consists of *self-organizing systems*, which are wholes of interrelated components. Obviously the *systems philosophy* is a description of the activity of reality, without implying existence. Also it fails to unify the diverse empirical systems.

In contrast, *system philosophy* is defined here as an integrative thought about the universe as a system of matter and consciousness, where these constituents are in dialectical and productive relation. It effectively shows that things exist by the union of opposites. This philosophical perspective about real existence involves the synthesis of rational and empirical aspects of knowledge.

Reading and reflecting upon concerned books, I enquired how to fit the theory of *self-organizing systems* within the framework of philosophical thought. A brief record of my quest is separately given under the forthcoming *Prologue*.

When we make an overview of the existing books in hitherto history of philosophy, as developed in the previous 3000 years, we can see two groups of philosophical books. The first group consists of books written by original philosophers who presented their own thoughts regarding specific areas of philosophy. The books of Plato, Aristotle, Descartes, Hume, Kant, Hegel and Whitehead are the prominent examples. These works are the source of academic arguments by professors and scholars in the field; but students and ordinary people find it extremely cumbersome to understand.

In the second group we consider the books prepared by scholars who try to clarify and explain the works of original philosophers. Most of such secondary books have focused on the doctrines of individual philosophers in a chronological perspective; but they made very little effort to evaluate and integrate the apparently diverse lines of thought. In other words, they do not deal with the philosophical *themes* in a coherent manner.

Can we synthesize these divergent streams of philosophy? This is the central question of my present book Discovery of Reality. All chapters have a considerable set of original ideas, clearly marked by the symbol [*]. The main examples of **innovative ideas** are listed below chapter wise.

Chapter 1 -- *A Guide to the Levels of Knowledge.* The first chapter introduces systematically the methodological stages called Theory, Hypothesis, Deduction, Testing and Induction, which are abbreviated as Ty, H, D, T and I respectively. The distinction between content view and process view as well as the organic levels of knowledge has important application throughout the book. The 2x2 Table -- of content view, process view, rational view and empirical view -- is introduced innovatively.

Chapter 2 -- *Functional Classification of Philosophy.* At the outset, a new *definition* is advanced - philosophy is an objective and integrative knowledge for reconciling the diverse paths of science, religion and art. Then core functions and objectives of Philosophy are explained under two heads - ontology (theory of reality) and epistemology (theory of knowledge).

Chapter 3 – *Worldviews and Versions of Reality.* Here we will focus on western philosophy only. In the early stage, Greek philosophers proposed the opposite theories about *mind* and *matter* as the fundamental stuff or reality of universe. Subsequently, various conceptions about mind and matter emerged, which can be analyzed using the notion of **worldview** that is defined here in an original manner. Thus there are six worldviews adopted in the history of western thought under the main groups of organic worldview, spiritual worldview, mechanistic worldview and physical process worldview. It will logically lead to seven theories of reality. Then the six worldviews would serve as the framework for giving the *basic classification* of the entire range of diverse philosophical doctrines.

Chapter 4 – *Western Philosophy - Key Doctrines of Great Philosophers.* This chapter is devoted for a detailed analysis of the main areas of western philosophy as given below.

- Idealism (mainly the doctrines of Socrates, Plato, Aristotle, Augustine, Thomas Aquinas)
- Spiritual process philosophy such as pantheism, historicism and evolutionary theology (Heraclitus, Zeno, Plotinus, Spinoza, Leibniz, Hegel, Teilhard Chardin and Whitehead)
- Deism and Rationalism (Descartes and Immanuel Kant)
- Materialism and Naturalism (John Locke, George Berkeley and David Hume)
- Mysticism, Phenomenology and Existentialism as well as the Physical Process Doctrines like analytic philosophy and linguistic analysis.

This description of prominent doctrines of western philosophy would pave the way for the introduction of System Philosophy.

Chapter 5 -- *System Philosophy of Reality.* This core chapter defines *reality (ultimate reality)* in a comprehensive way. The points of originality are mentioned below:

a) The history of diverse *theories of reality* is explained by introducing the phrase *monistic philosophy*; it distinguishes the erstwhile philosophy from the newly proposed System Philosophy.

b) Then we establish that, based on *six worldviews*, western monistic philosophy has proposed seven ontological divisions. Through our synthesis, we get the definition: *Ultimate Reality is the system of opposite forces called Body and Consciousness, which are represented by X-axis and Y-axis respectively;* it is called the **System Model of Ultimate Reality**.

c) Then reality is like a factory having opposite components of infinite measure. This model establishes the existence of seven social systems at the global level into which human life is divided. Importantly, it leads to the solution of the longstanding *problem of evil*.

d) Further the logical problems of various religious doctrines – like theism and pantheism – are solved. Similarly, the Brahman-Maya doctrine, under the Vedanta philosophy of Hinduism, is interpreted in a unique manner. The system model illustrates Maya as the X-axis while Brahman is the Y-axis, implying complementary existence.

Chapter 6 -- Seven Life Systems and Knowledge. In this chapter, we introduce a unique way of classifying the numerous social institutions, which preexist for imposing the values and goals to social relationships. This leads to the innovative idea of **seven life systems** for referring to non-overlapping institutions at the global level. They exist by the dualistic relation between self-interest (SEI) and society-interest (SOI) pertaining to various faculties of human mind. Accordingly, the system model will pave the way for the epistemological synthesis of the entire spectrum of human knowledge.

Chapter 7 -- Comprehensive View of Truth. We define the problem of truth by noting the dilemma associated with Cartesian Theater. For proposing the *comprehensive view about Truth*, we have to unify the truths of various branches of knowledge. In the light of System Philosophy, we can hold that every kind of knowledge (TyHDTI scheme) has **necessary-contingent truth**. This innovative theory can be illustrated by the *system model* and it removes the dichotomy between *necessary truth* and *contingent truth* in the propositions. The central problem of ethics -- definition of good and bad -- is solved now. Also we get the method for linking fact with value. The ontological existence of truth and falsehood, with various levels, is demonstrated using the system model.

As per the foregoing, my endeavor is to give a reply to the doubting people who ask: what is the use of philosophy? This book is written in a pedagogic style, by explaining all concepts in a basic manner making use of suitable Tables and Diagrams. So the book would serve as text as well as reference for all persons, who desire to know the answers for questions about the basic aspects of any discipline.

George Luke

PROLOGUE

MY EXPERIMENTS WITH PHILOSOPHY

My life story has many unexpected twists and turns, which reveals the specific intentions of *almighty, the reality of universe*. This brief autobiographical note is intended to reveal the trials and tribulations, which I experienced in the course of my philosophical quest. I was born on 03 June 1953 as the eldest son of Aleykutty and P. L. George belonging to Puthankulam family of Thodupuzha Taluk in Kerala state, south India. Father was a teacher of Malayalam language. I am a member of Roman Catholic Church of Christianity.

Starting life in an underdeveloped part of Kerala state, I studied in Malayalam medium schools - at Neyyassery (1958–62) up to fourth standard and afterwards at Kodikulam (1962-68). In the primary and middle levels I was very poor in mathematics. But during the year of eighth standard I had a sudden interest in Geometry, particularly in proving theorems. It kindled my aptitude for mathematics, which in coming years helped to expand my brain power. I passed SSLC in 1968 meritoriously, having placed at fiftieth rank in the all-Kerala list. I received the national merit scholarship also.

During the vacation after SSLC examination, I started reading books available in local library. Having good proficiency in reading and writing essays, I curiously read books on popular science including the articles that appeared in periodicals like Mathrubhumi weekly, especially relating to Quantum Physics and Cosmology. This interest in scientific topics prompted me to think on the deeper aspects of world and to brood over philosophical questions. My mathematical and logical mind became suitable for creative thinking on the ultimate issues, though I did not have formal education in philosophy.

I Graduated from Newman College (1968-73) at Thodupuzha, securing second rank in B.Sc. (Mathematics Main) of 1973 Batch of Kerala University. It is worth mentioning that I got hundred percent marks in all Mathematics papers. My postgraduate education was in the Department of Statistics, Kariavattam Campus, University of Kerala, Thiruvananthapuram. I secured first rank in M.Sc (Statistics) of 1975 Batch. Though there was an offer of lectureship from a good college, I was interested in higher studies; hence continued in the same Department for newly started M.Phil (Statistics) course and came out in next year with A Grade.

By that time, having studied econometrics in the post graduation level, I was pondering to do research in some practical problem related economics. Keeping this aim in view and due to other circumstances, I joined Reserve Bank of India in Mumbai on 28-09-1976 as Statistical Assistant and became Staff Officer on 28-04-1982. Then I switched over to National Bank for Agriculture and Rural Development (NABARD) and served as Assistant Manager at its Mumbai Head Office during 1983-85. With next promotion to the post of Manager, I was transferred to the regional office at Thiruvananthapuram (1985-1993) and later to Pune (1993-2001). During these twenty-five years, I could acquire professional experience by conducting various studies in connection with official work including inspection of cooperative banks.

While working in Mumbai, my intellectual life was immersed in banking and other fields of economics. I tried to do part-time research in economic department of the University of Mumbai; but my request was turned down due to lack of M. A. degree in economics. This prompted me to conduct self-study of economic texts; but I could not write exam for post graduation in this subject on account of other priorities.

Reaching Thiruvananthapuram, I continued my struggle with serious subjects alongside official work. I can recall that the seed of my interest in philosophy was sown by a special incident in 1988. On a Sunday morning, I was electrocuted while operating the washing machine in my home. Lying in water with electricity flowing through my body, unable to move or speak, I was expecting the imminent death. At that time my co-brother, who was in the next room, came and saw my danger; he rushed to put off the main switch, thus saved my life. This near-death-experience caused churning in my mind and death became a frequent subject for thought. Subsequently, the aptitude for learning philosophy had been growing in the forthcoming years. I used to get shivering due to wonder over the thought that the universe is extended infinitely without end.

Since my temperament was tuned to research activities, the job related to banking appeared to be quite routine and uninspiring. Taking books from the great libraries of British Council and Center for Development Studies in Thiruvananthapuram, I attained considerable depth in various topics of economics. The production function model of classical and neoclassical economics was the center of my intellectual fascination.

During the last period of my stay in the city, my ambition of part-time research in economics was revived. With the guidance of Professor M. A. Oommen, I prepared an article titled *Productivity of Capital Investment on Marine Fishing Crafts of Kerala* and got it published in July-Sept. 1993 issue of *Productivity*, Journal of the National Productivity council, New Delhi. On the basis of this article, I was selected in February, 1993 for research in Dept. of Economics at Kariavattam Campus of the University of Kerala. But the PhD program could not be started due to the immediate transfer of my official posting to Pune, which is reputed for better academic atmosphere. After reaching Pune, I could get the support of a guide for part-time economic research in the Mahe center of Pondicherry University, but I was denied permission on account of the lack of concerned post graduation.

Major highlight of my thought process, in the next phase, is the deviation from the ongoing study of economic themes. As a tryst with destiny, in November 1994, I purchased the popular book of Fritjof Capra, *The Turning Point,* which was originally published in 1982. Here Capra describes various layers of world – subatomic phenomena, biological organisms and different social organizations – as evolving systems, which are wholes of interconnected parts. Then he resorts to Chinese mysticism for addressing the ultimate questions. Since a whole is more than the sum of parts, it has holistic, ecological and dynamic existence. Accordingly, Capra's *systems view* of life and economy appeared to counter the mechanistic approach of the corresponding disciplines.

The treatment of systems in Capra's book triggered my critical mind and it became the turning point in my intellectual pursuit. I recognized that Capra has not explained the origin of life in the inanimate macromolecules like DNA. And, he failed to present a philosophical perspective about systems. What is the philosophy about the interconnected but layered world? How can we compare mysticism with the alternative thoughts about reality? What are the essential principles behind capitalism and communism? *These questions agitated me and I decided to take up the study of philosophy as a new venture.* I enlarged my reading by purchasing hundreds of useful books in philosophy, social sciences and related subjects from exhibitions and book stalls.

The persistence of poverty and high inequality of income-asset distribution had been a matter of grievous concern to me, in the course of analyzing the economic problems. Another issue that pained me was the cruelty and destruction due to wars as well as terrorism. Why do political leaders commit such atrocities, even though religions profess love and peace? Pondering over these problems I came to the conclusion that most of the crimes and evils are performed at the social level, rather than at individual level. The main drawback of idealism and religious philosophy is its focus on individual mind without giving due importance to the patterns of social behavior.

Thus I began to use the social perspective for deliberating about world. The most important challenge for me was to explain that social systems exist by the complementary relation of opposites. Matter-energy, space-time, body-mind, self-society, capital-labour are prominent examples of opposites, when we consider various phenomena. Continuous exposure to my books and thinking over different aspects of world generated a unique idea in me that our life is spent in seven global social systems; this principle is denoted by the phrase *seven life systems*. It became the key to start my philosophical project, which was named *System Philosophy*.

In this context, I got an intuitive idea that the *production function model* of economics would give an innovative method to study the interconnected behavior of opposite entities as well as social systems. However, for articulating this proposal, I have to study philosophical doctrines seriously. Gradually I realized that the routine of official job is a hindrance to my progress in this direction. Moreover, I was suffering from diabetes for some years on account of the continuous strain of banking career and private study. Due to the pressure of such circumstances, I voluntarily retired from NABARD service on 29-9-2001 for engaging with the research, writing and publication in the field of philosophy.

Sitting at home, I plunged into the selected books in my possession for developing the themes of System Philosophy. The dialectical method that I resorted was to take notes in English and then translate it into the mother tongue Malayalam. Then the reverse method also was adopted. This bilingual process has helped me to increase the clarity of philosophical issues.

My deep interest in the production function model of economics generated a novel idea in me for depicting the reality of universe as well as our social systems. It is the first time that a mathematical model is employed in philosophy for explaining its abstract concepts.

Inspired by the hope to articulate System Philosophy in an innovative manner, I engaged myself for three years in preparing the manuscript of my first book in Malayalam. Then I approached certain important publishers, but they refused to publish my philosophical book, holding that it is difficult to sell in current scenario. So I self-published the book in April 2004 with the title **Saptaloka Darshanam**

Samgraham *(philosophy of seven life systems – a summary)* under the banner of PGL Books, Changanacherry, Kerala – 686 101. This work was mainly intended to explain the theoretical concepts of the seven global systems namely *nature, economy, politics, family, ethics, religion* and *art*. Due to the popular and traditional importance of number *seven*, many readers were amazed at my classification.

Next development is my participation during 2004-2005 in the program of the School of People's Economics, conducted by the NGO called VICHARA at Mavelikara, Kerala. It consisted of about thirty days of discussions and seminars on various topics, which sharpened my philosophical ideas. As a result, I prepared an article titled *From Modern Science to System Philosophy* and published it in the June 2005 issue of the journal OMEGA of ISR Aluva, Kerala.

As a matter of divine providence, I got the opportunity to join a research program during 2005-2008 under the Association of Science, Society and Religion (ASSR) of Jnanadeep Vidya Peedh (Papel Seminary), Pune. The discussions and seminars conducted in December month of these years as well as the library facilities helped me to research on the interface between science, religion and philosophy. I presented a dissertation on this topic and it became the spring board for my intensive pursuit in the following years. So far I have purchased over thousand academic books that serve as authentic references for full time creative work, while remaining in the modest facilities of home.

- I presented a paper titled *Whitehead and particle-wave duality: A critical appraisal from the perspective of System Philosophy* in the 7th International Whitehead Conference, January 5–9, 2009 held at Dharmaram College, Bangalore.
- In March 2015, I self-published an authentic book ***jeevanum parinamavum*** (Life and Evolution) in Malayalam language discussing the theories of biological phenomena in the light of *System Philosophy*.
- I presented weekly Radio talk on various subjects regularly for six months in 2015-16.

- I have some letters to Editor regarding social issues published in news papers and periodicals of Malayalam language.
- I got an article titled *Ayurvedathinte Jnana Sithantham* (Theory of Knowledge of Ayurveda) published in the May &June 2017 issues of OUSHADHAM Journal of Ayurvedic Medicine Manufacturers Organization of India (AMMOI), Thrissur, Kerala.
- Started in 2006 a website *www.systemphilosophy.com* for presenting the topics of System Philosophy in simple English and Malayalam. I have posted a few important articles in this website.
- Further, recently I have some posts pertaining to philosophical ideas in my account at *www.facebook.com/LukeGeorge*.
- Organized the *Academy of System Philosophy* under a trust to disseminate the new philosophy among wider audience.
- I regularly participate in seminars and discussions concerned with science, religion, philosophy and various social issues.

The foregoing is the background for preparing the manuscript of the comprehensive book titled *Discovery of Reality : The Light of System Philosophy*. In the recent years I have exchanged ideas with many well placed scholars; but to my surprise they are finding it extremely hard to understand the principle of system as elaborated in my writings. Generally people are obsessed by the thought that opposite entities are separate, without any interconnection.

Here I may mention the fact that acquiring knowledge is a social process, which is influenced by the ideologies and vested interests of powerful individuals of society. If a new idea comes from a lover of wisdom, who lacks the support of institutions like universities and media, it will normally face the struggle for existence. I believe that the esteemed readers of my book will help its natural selection in future because human mind has an innate tendency to prefer truth and discard falsehood.

Many teachers, friends, relatives and well wishers have helped me in the course of my life so as to contribute to the evolution of my philosophic views. My deep gratitude to all of them is beyond words. This book is dedicated to my parents and other ancestors from whom I inherited the genes. I am especially indebted to many writers of philosophy and related subjects, as mentioned in the notes of the chapters of this volume. I can emphasize that my limited words are not sufficient to express my acknowledgement of the ideas received from the forerunners in the history of thought.

George Luke
03--06--2018

INTRODUCTION

Discovery of Reality- The Light of System Philosophy is the third book in the series of three books dealing with the fundamental aspects of world and phenomena. The other volumes are *Origin of Universe* and *Life and Mind* As a combination it forms a comprehensive treatise on REALITY by George Luke who took voluntary retirement at the age of forty eight to prepare his *magnum opus*.

The author has shown that *System Philosophy* is an integrative thought about the universe as a system of matter and consciousness. This can be contrasted from the empirical and pluralist concept of *systems philosophy*, which has been popularized in the writings of Ervin Laszlo. The present book has been divided into seven chapters.

The first chapter is titled *A Guide to the Levels of Knowledge*. It introduces systematically the structure of knowledge, which is basically expressed as propositions. The author explains the way propositions are organized into separate methodological stages called Theory, Hypothesis, Deduction, Testing and Induction. Then he introduces the organic levels of knowledge; it has important application throughout the book.

Chapters 2 and 3 would provide a comprehensive account of the main divisions of western philosophy. For classifying the set of philosophical doctrines, the author introduces the concept of *world view* in the following words:

"We know that the theories of science, religion and art are based on our experiences of phenomena in different ways. Such theories can be treated as various phenomenal descriptions of reality of universe. Now we will explain the bridge between reality and the set of theories of phenomenal world using the notion of worldviews.

When we experience and study phenomenal world, we must take into account the relation between body and mind as appearing in inanimate and living things. The phenomenal descriptions about the existence of body and mind and their mutual relations become the fundamental ideas upon which the theories of science, religion and art are to be constructed. A particular conception about reality is philosopher's enterprise. Here we define worldview as the method for classifying the totality of theories in accordance with the perspectives of rational view, empirical view, content view and process view. Accordingly, a worldview is the set of common ideas found in the broadest family of theories. In other words, it contains the essential ideas of all theories having family resemblance".

Then the author shows that there are six worldviews in the history of western thought and consequently seven theories of reality. Thus he presents the basic classification of the entire range of diverse philosophical doctrines.

Further, chapter 4 is devoted to explain lucidly the basic teachings of great philosophers of western world. It clears the way for introducing System Philosophy.

Chapter 5 titled "World and Reality" is the CORE chapter of this volume, *Discovery of Reality*. The author makes a distinction between phenomenon and reality as follows: "We can define phenomenon as any object that depends on another object through cause-effect relationship. On the other hand, **reality is the original cause of all phenomena taken as a whole**. Accordingly, reality is self-caused, infinite and permanent. The terms like *ultimate reality* and *ultimate truth* are commonly used as synonyms of reality".

The bridge between reality and the set of theories of phenomenal world is established by the author using the notion of worldviews. He then introduces the **System Model of Ultimate Reality.** It is defined as the system of opposite forces called Body and Consciousness, which are represented by X-axis and Y-axis respectively. Then reality is like a factory having opposite components of infinite measure. This model can additionally explain the existence of seven manmade social systems at global level. Further, the above theory of reality serves to depict in philosophical manner the existence of God and Evil as polar opposites.

In Chapter 6, George Luke deals with Seven Life Systems and Knowledge. I find this quote of author interesting: "The spectrum of social knowledge must be first divided into *scientific social knowledge* and *mystic social knowledge*, in accordance with our faculties of scientific mind and mystic mind respectively". The names of the *seven life systems*, which are global-level social systems, are proposed as below:

1. Natural Life System (NLS),   2. Economic Life System (ELS),
3. Political Life System (PLS),   4. Family Life System (FLS)
5. Ethical Life System (ETLS),   6. Artistic Life System (ALS)
7. Religious Life System (RLS)

Chapter 7 discusses the philosophical dilemmas about truth. Author begins with the definition: "Truth is the property of a justified belief that it corresponds to an actual state of affairs of the universe". A new theory about *necessary-contingent truth* is the main highlight of this treatise. It is interesting to note that the system model of truth unifies the different kinds of truth in various disciplines.

Finally the author hinds at the further path of System Philosophy through extending to the study of social sciences like economics, politics and ethics. The principal aim of author is to link fact with value. Thus, adopting system approach, the author claims to solve the age-old conceptual problems of physical, biological and social worlds.

George Luke has done a commendable job in preparing three volumes - Origin *of Universe, Life and Mind* and *Discovery of Reality* - using concepts of System Philosophy. I congratulate him for this singular achievement.

Hardev Singh Virk
Professor of Eminence
Punjabi University
Patiala, Punjab (India)
www.drhsvirk.weebly.com

14 February, 2018

Chapter 1

A Guide to the Levels of Knowledge

****

1.1 Definition of Knowledge and its Classification

1.2 Organic Levels of Propositions

1.3 Approach to Theory of Knowledge

Author's main original ideas are marked by [*].

The mark [#] gives the number of note at the end.

****

In the wake of the tremendous expansion of knowledge in the recent centuries, it is remarkable that the gulf between scientific ideas and religious faith also widened. This dispute is serious when we consider the questions: How did the universe originate? What are the definitions of matter, life and mind? Does the world have purpose? Is the world created by God? Does God exist? Of course, science and religion are prone to advance divergent answers to these questions. Additionally we have the experiences of artistic beauty and it produces certain kind of knowledge distinctively called as *art*. This is the proper setting for initiating an enquiry into the basic features of various kinds of knowledge, mainly science, religion and art. [# 1][*].

The systematic study of knowledge is concerned with the following questions:

(q1) What is knowledge?
(q2) What is the *method* of forming the constituents of knowledge?
(q3) What are the *faculties of human mind* (source) used for acquiring different kinds of knowledge?
(q4) How can we *justify* particular piece knowledge? Is there a common criterion of justification applicable to diverse situations?
(q5) What is the meaning of *truth* in respect of knowledge?

This chapter gives only basic and introductory points about these important questions. The deeper issues are to be explained in the forthcoming chapter pertaining to the doctrines of philosophy. This would prepare the ground for further deliberations in this book. [# 2][*].

1.1 Definition of Knowledge and its Classification

At the outset we hold that knowledge is a set of ideas possessed by a person; such knowledge is stored in the books, news papers and magazines as well as other means like electronic devises. In this situation, it is a commonsensical statement that our knowledge is expressed through the medium of certain language. The most elementary components of a language are *alphabets*. When a set of alphabets is given particular meaning it is called a *word*. A group of words are arranged according to the laws of grammar to form a *sentence*. The essential property of sentence is that it contains one or more ideas. Hence, we can define sentence as a combination of words for expressing some idea through the medium of language. Obviously, the sentences with same idea are different for separate languages like English, Malayalam and Tamil.

Sentences generally have two categories. First, the group of assertive sentences expressing some idea about objects or events in the world. Second, the set of non-assertive sentences denoting questions, commands and exclamations. See the following examples.

1. Question: What is the time now? What is your name?
2. Command: Shut the door. Get out.
3. Exclamation: Oh, what a heat!

Here we are concerned with the *assertive sentences* pertaining to some object or event in the universe. These sentences have meaning as they express the idea (information) about the existence of some state of affairs. It is expedient to divide the range of assertive sentences into two classes – subjective and objective. A subjective sentence expresses the feelings of a person including all kinds of emotions, imaginations, opinions and illusions. It is easy to recognize that various forms of art and literature such as poetry, drama and novel fall within the realm of subjective sentences. Suppose the author of a novel describes a character by saying that he was wearing white shirt in an occasion; it is clearly an expression of the imagination of the author.

Now we may define *proposition* as an objective sentence which is valid on the basis of a set of evidences. As an elaboration, a proposition possesses the following three essential characteristics:

(c1) *Proposition asserts, on the basis of evidences, the existence of certain object or event in this universe.*
(c2) *The information contained in the proposition may be true or false.*
(c3) *Proposition is a belief expressed in third person perspective.*

For beginning the discussion about knowledge, it is customary to distinguish between the two notions: knowing *how* and knowing *that*. The former pertains to knowing how to do an activity, such as knowing to drive a car, knowing a person and knowing a place. The latter kind of knowledge is expressed by way of propositions as defined above. Each proposition is an assertion in the form "S know that p", where S stands for the subject or person having the knowledge indicated by *p*. It may be added that we can make propositions about our knowing how also. Hence, in the context of philosophical study we have to focus on propositional knowledge only.

The first, second and third perspectives are aspects of grammar since these are basically linked to the words I, You and It. We can add that he, she, they, that, thing and similar words stand for 'it' depending on the context. Accordingly, 'earth is round' is a proposition – it is expressed in third person perspective referring to a particular planet of sun. Similarly, the sentence 'yesterday I had a dream of walking upon Mars' also is a proposition because it expresses the occurrence of a dream. The content of the dream is a subjective experience and it is not relevant here. The said proposition is true if the event of dream is established by factual evidences.

Based on the above analysis of propositions we can define knowledge as following: ***knowledge is a true proposition***. It gives correct information about the aspects of an event or object on the basis of evidences. We can exclude the subjective sentences belonging to art and literature from the field of knowledge for reasons mentioned above. However, it is possible to analyse and criticise such sentences objectively; the topics of literary criticism is a mainstream example. In this manner knowledge is possible in the form of propositions pertaining to the features of our subjective experiences. The religious sentences in first person perspective such as 'God exists' and 'God is love' can also be converted into objective mode. Then we can deliberate whether the propositions are true in accordance with publicly known facts or evidences.

Our deliberation about knowledge starts from the view that knowledge is the product of the mind of a person or a group of persons. This point is underlying the condition that *proposition is a belief that is a mental state*. It is necessary to keep the human mind at the center of the landscape of knowledge, for embarking on our philosophical contemplation. When we hold that propositional knowledge is a belief, it means the following:

> ➤ Knowledge is produced in a particular faculty of mind on the basis of aptitude and purpose of the person. Such knowledge can be stored in a book or other medium.

> The mental state for making propositional knowledge is closely associated with a set of feelings and emotions; hence it prompts for action in specific cases. This implies that knowing is an active function of mind, not a passive one.
> For a person to get knowledge, he or she must have desire for it. For example, suppose two persons A and B are engaged in conversation. A is trying to explain a particular philosophical subject to B. If B is not attentive, lacking the motivation to understand philosophy, the talk of A would not generate belief (knowledge) in the mind of B.

Can we distinguish strictly between knowledge and faith? The ideas of science and religion both are constructions of our knowing mind. This leads us to admit that science is a separate form of faith adopted by people following the so called scientific method. This view enables us to hold that different subjects like science, religion, art and philosophy are separate levels of knowledge (faith). This unifying approach is necessary for the analysis of knowledge to be undertaken in this chapter. In the following paragraphs, we will examine the categories of different disciplines of knowledge. For that purpose, it is necessary to consider knowledge in two stages – first stage deals with the level of propositions and the second stage takes into account the groups of propositions arranged in the order of higher disciplines.

Classification of Propositions

As explained above, the unit of knowledge is a true proposition. In this situation our discussion about the features of knowledge must start with the taxonomy of propositions; it aims to facilitate the objective understanding of knowledge. Since knowledge is the creation of our mind, it is reasonable to connect the taxonomy with the fundamental aspects of our mind. The following four methods of classification of knowledge are crucially important for the pursuit of this book.

a) Fact and Value

Here we may propose that *fact* and *value* represent the dual purposes of our knowing mind. The first purpose is to know about the existence or characteristics of the concerned object as well as the circumstances causing change in the object. In other words, the opposite aspects of permanence and change are considered in the stage of acquiring knowledge – the resulting propositions are collectively called as *facts*.

The second purpose of knowing mind is to assess the utility of object as far as our life is concerned. Various kinds of information about such utility maybe denoted generally as *value*. Obviously, value represents our objective to recognize the qualities of the object that are good or bad. The examples of good values are love and courage. On the other hand the words such as hate, timidity, sadness, denote bad values depending on the situation concerned. Since value represents the intentions of our mind, it has positive and negative directions. In other words, values appear as duals like good-bad, beauty-ugly, love-hate and so on.

b) Content View and Process View

The descriptions about the structure of an object are often expressed in content view. It deals with the static existence of the object having such and such components. The words denoting the constituents of the object have representational contents. For example, consider the proposition 'earth is round'. Here 'earth' is a word under content view since it implies that earth exists statically as an object with certain perceivable components. Additionally, the word 'round' represents the geometrical object called sphere. The definition of things as appearing in a dictionary is produced under content view. Further, for example, the laws of classical science and religious knowledge as well as ordinary perception of world also belong to this category.

Process view is concerned with the description about the circumstances causing change in the object. Change can be alternatively viewed as an activity or process. For instance, the growth of a plant is the aspect of change happening to it; the same can be treated as the activity of plant. In the case of the growth of plant it is caused by certain circumstances such as the biological features of plant as well as the external factors like availability of water, manure and proper climate. The word *context* is generally used to refer to the totality of internal and external circumstances causing change.

It may be reiterated that the process view about an object consists of the descriptions of the related context causing change. The object itself is not considered for study. When we describe the context for the growth of a plant there may be many plants satisfying such conditions. This point is better expressed by saying that the context has *multiple realisability*. To put it simply, process view aims to describe the context without considering the concerned object. Main examples of knowledge under this view include quantum physics, mysticism and theory of evolution.

We have already mentioned the two classes of propositions pertaining to the permanence and change of an object. These opposite features are complimentary. Every object has permanence or static existence for a particular period of time. And it undergoes change when we consider the static aspects of different periods. This pen has permanent existence today, but it may be destroyed tomorrow due to some circumstances. The point which we stress here is that the knowledge about an object consists of both content view and process view, both of which can be applied to fact and value.

c) Rational and Empirical

It is well known that the typical examples of rational propositions are those belonging to mathematics and logic. Additionally we can include the various abstract concepts about things of universe. Generally speaking the class of rational propositions is further divided into definitions, meanings of words as given in a dictionary, concepts about quantitative measurements and qualitative values as well as the cause-effect relation. More often, we will call this kind as **deductive propositions**.

On the other hand, the class of empirical propositions – alternatively called as **inductive propositions** -- pertains to various kinds of knowledge obtained through our sense organs, namely, eye, ear, nose, tongue and skin. When I see this table I get the idea that its length is six feet and breadth is four feet. The propositions about length and breadth are obviously empirical, since it is derived from particular sensory perceptions. It may be added that the empirical propositions are constituted using the rational ideas also in the form of definitions and other abstract concepts. In this example, the words like table, length, breadth, six, four and feet denote rational definitions. Here, we have given only an introduction of the distinction between rational and empirical propositions anticipating detailed treatment later.

d) Simple and Complex Propositions

A proposition holding a single idea or information is characterized as simple. Examples: this is a table, matter has weight and God is love. The definitions of numbers in mathematics are also simple propositions. When we combine two or more propositions as per the rules of grammar it becomes a complex proposition. Example: chair is smaller than table. This is a complex proposition since it includes the definitions of table and chair as well as the comparison between their sizes. The methods of logic and mathematics are used for combining various propositions to form more complex ones. The proposition $2 + 4 = 6$ and all other equations of mathematics are typical examples of complex propositions.

The propositions of both fact and value can be classified into simple and complex propositions. Then it is possible to divide the total corpus of human knowledge into various disciplines, which constitute certain hierarchical levels. However, I introduce here the innovative idea that the innumerable topics or disciplines can be finally brought to a 2x2 Table using the criteria of Content View, Process View, Rational and Empirical.

This method of 2x2 classifications would become highly effective for reconciling the controversies under *theory of knowledge*. It may be anticipated that the study of any topic – for example: matter, mind, God, soul, light, sound, life or evolution – must be conducted from the framework of this Table.

Table 1: Broadest Classification of Disciplines [# 3][*].

	Content view	**Process view**
Rational view	*(List of* Disciplines)	*(List of* Disciplines)
Empirical view	*(List of* Disciplines)	*(List of* Disciplines)

1.2 The Organic Levels of Propositions

From the taxonomy of propositions just presented we can derive the order in which human knowledge has expanded. Considering the analogy of a tree, we observe that the *tree of knowledge* has grown vastly with countless branches and leaves; and it is still growing every moment. A cursory glance at the large spectrum of subjects taught in the universities includes such diverse disciplines as arts, science, medicine, engineering, social science, theology and so on. These are the branches of the tree of knowledge; the proliferation of sub-disciplines point to the miraculous expansion of our cognitive mind. We can present a systematic outline of the tree of knowledge with the help of biological evolution as a model. In this way we want to see the landscape of human knowledge for identifying the relative positions of various disciplines.

Also it reveals the hierarchy of complexity achieved from the basic levels of propositions.[# 4][*].

Biology adopts the classificatory scheme for living world in the order of cell, organs, organisms, species, family, class, phylum and kingdom. Thus at the highest level, the living beings are divided into plant kingdom and animal kingdom. The same method can be adopted for presenting the order of disciplines in the world of knowledge. The simple propositions, which are the basic units of knowledge, are like cells. Combining a few simple propositions in a meaningful way we produce complex propositions – these are similar to the multi-cellular microorganisms. Then higher levels of complex propositions are subsequently produced.

Here we may identify that the most important kind of knowledge is the inferences about cause-effect relations or laws. "Cloud causes rain", "God is love" and "Artistic beauty provides pleasure" are examples of various laws, respectively under science, religion and art. It has been explained earlier that such laws or cause-effect relations are produced by a sequence of stages called Theory, Hypothesis, Deduction, Testing and Induction – these stages are abbreviated as Ty, H, D, T and I respectively. These higher groups of propositions can be compared to the organs of an organism. It may be emphasized that knowledge about an object or event is constructed by organizing propositions in such five stages.

What is explained through simple examples is that knowledge about a particular thing or law essentially consists of various propositions that are to be grouped into the five methodological stages. The fifth stage called inference is the proper knowledge about thing or law. In the context of science such inferences can be alternatively called as physical laws. It is convenient to use the word 'laws' for referring to all kinds of scientific inferences pertaining to the external world.

In the world of knowledge, each law is an individual being that can be compared to an organism of biological world. In other words, the laws are the separate individuals in the knowledge world; each law is constituted by the organs denoted as Ty, H, D, T and I. Alternatively, we can hold that propositions of five methodological stages combine to form a law (organism). Now it is possible to identify the higher groupings of laws by adopting the method of biological classification. For that purpose the dual properties of similarity and difference are taken into account in respect of the countless laws in the realm of knowledge. Those laws which are more similar are based on a particular theory. Such laws can be grouped into a subject that is similar to a **species**. For example, electricity is a subject (species) which includes a group of interrelated laws of physics. The following **Table 2** shows the higher levels of classification in the range of subjects.

The macro fields of knowledge -- mainly science, religion, art and spiritual science -- give direct knowledge of universe on the basis of different kinds of experiences. It may be clarified that such direct knowledge of things and laws are obtained through our sensory experiences joined with abstract thinking. From this perspective, the knowledge of science, religion, art and spiritual science together is characterized as *first order knowledge*. In contrast, **philosophy is to be described as *second order knowledge* because it aims to synthesize the various kinds of first order knowledge.** The primary endeavor of philosophy is to analyze the various concepts used for knowing the universe in diverse manner like physical, religious and aesthetic ways.

Table 2 : Organic Levels of Knowledge [# 5][*]

	Biological World	Knowledge World
8	Kingdom	**Macro fields** of knowledge– science, religion, art, philosophy and spiritual science.
7	Phylum	**Division** – physical sciences, biological sciences, social sciences, humanities, theology, etc.
6	Class	**Subject class** – physics, chemistry, biology, economics, and so on.
5	Family	**Subject groups** – astronomy, classical physics, quantum physics, and so on.
4	Species	**Subjects or disciplines** – electricity, sound, light, and so on.
3	Organisms	**Laws** - inductive inferences in various subjects.
2	Organs	**Methodological stages** -- Ty, H, D, T and I
1	Cells	**Propositions**

1.3 Approach to Theory of Knowledge

Now we will consider the task of philosophy for unifying the various kinds of first order knowledge. The branch of philosophy taking up the study of general aspects of knowledge is named as *epistemology* or *theory of knowledge*. Out of the five questions mentioned in the beginning of this chapter, we have so far dealt with the first one: what is knowledge? This is the stepping stone to the other questions denoted by q2, q3, q4 and q5, which pertain respectively to *methodology, source, justification and truth*. These aspects together constitute the proper area of epistemology.

Through the pursuit of epistemology, we want to develop a vision about the area of *ontology*, which describes the original cause or reality of universe. A systematic description of these main functions of philosophy -- *epistemology* and *ontology* -- is postponed to next chapter.

NOTES of Chapter 1

1   It is my original idea to initiate the study of the basic features of knowledge by referring to the science-religion conflict. This approach opens the gateway to the enquiry about core areas of philosophy in this book.

#2   Main reference books used for this chapter are: Anthony Harrison– Barbet (1990), Brooke Noel Moore and Kenneth Bruder (2005), Chatterjee (1988), Copleston (1994), Ewing (1994), George Thomas White Patrick (1978), Grayling (Editor) (1995), Grayling (Editor) (1998), Hospers (1997), Lavin (1989), Tarnas (1991), Thilly (2000), Thomson (1997), Urmson J.O. and Johanthan Ree (1989) and Washburn (1997).

The treatise of this chapter is built upon the points extracted from my reference books; and it is written in accordance with the frame work of System Philosophy to be developed in due course. The originality of this presentation will be evident if it is compared to the relevant sections of other books.

3   This classification of disciplines is my original idea.

4   The adoption of the pattern of evolution would enable us to remove most of the confusions in theory of knowledge.

5   Note that the mention of disciplines in this Table is illustrative, not exhaustive.

Chapter 2

Functional Classification of Philosophy

****

2.1 Objectives of Philosophy

2.2 Ontology

2.3 Epistemology

Author's main original ideas are marked by [*].

The mark [#] gives the number of note at the end

****

A systematic introduction of the main streams of philosophy is vitally important for the future course of this book. It is popularly known that the aim of philosophy is to answer the ultimate questions regarding the universe as well as the human experience. A few samples from the list of issues addressed by this discipline are:

- What is the reality of universe underlying the various kinds of our experiences?
- Does the universe have purpose?
- Is there a God as the creator of universe?
- What is the mind? What is the soul?

- How do we get different kinds of knowledge like science, religion and art?
- What is truth? What is beauty? What is justice?
- How can we distinguish between good and bad?

The curious feature of the philosophical subject is that such questions are not answered in a definite way; there are different schools of thought dealing with a problem from separate points of view. More specifically, there is considerable variation between the philosophers with regard to their concepts, observations and method of interpretation. In this predicament, the desire to compare and evaluate the competitive traditions is uppermost in every student of philosophy. However, no progress has been achieved toward this end for want of an appropriate framework to conduct a comparative study. But the situation can be changed radically through the analysis of philosophy to be presented in the ensuing sections.

It may be recalled that a good number of ideas and themes of philosophy have been introduced in the previous books of this author – *Origin of Universe* and *Life and Mind*. The present chapter aims to present the details of such doctrines as well as additional principles in the appropriate places so that the structure of philosophy can be explained.

We can see two groups of authors dealing with philosophical ideas. The first group consists of original philosophers who presented their own thoughts regarding specific areas of philosophy. In the second group we consider the writers trying to clarify and explain the works of original philosophers. Most of such secondary authors have focused on the doctrines of individual philosophers in a chronological perspective; but they made very little effort to evaluate and integrate the apparently diverse lines of thought. In contrast, the present chapter approaches the themes of philosophy with the aim of analysis and synthesis.

The best approach is to compare the discipline of philosophy to a group of trees, where each tree has certain main branches and further divisions. But the perception of these individual trees is based on the unifying concept of *tree*; we are fundamentally interested in the structure of a typical tree in an objective manner. Likewise, we may start with a comprehensive account of the core functions of philosophy. Then we proceed to identify the main branches of philosophy, named as ontology and epistemology, which are the theories about reality and knowledge respectively.

The most revealing aspect of our treatise is that the disagreements in the field of philosophy are on account of the divisions of ontology (theory of reality) as well as epistemology. For understanding the existence of various doctrines of philosophy, we will introduce the concept of *worldview* with a specific and innovative meaning. The salient features of these worldviews would serve as the framework for surveying the forest of diverse philosophical doctrines. Further, the topics of the subsequent chapters are to be analyzed using this structure of philosophical thought.

2.1 Objectives of Philosophy

We may start by clarifying the distinction between the terms called *reality* and *phenomenon*. It can be observed that the various objects of this universe are connected through an infinite network of cause-effect relations. The astronomical bodies like earth, sun and other stars are constituted by the particles of matter which existed in the earlier stage of universe. The formation of everything around us is accounted in terms of preexisting objects. According to this perspective, the conceivable universe of past, present and future consists of objects that are caused by previously existing objects; each object of this kind is called a *phenomenon*. The technical definition of phenomenon is: if object B is caused by preexisting object A, then B is treated as a phenomenon.

When we consider the causes of phenomena in a sequence towards the direction of past, we can reach at the notion of original cause. **Then reality is defined as the original cause of all phenomena taken as a whole.** Accordingly, we can hold that reality is self-caused, permanent and infinite. Considering the finiteness of every phenomenon, it is reasonable to conceive that all phenomena in this universe are ultimately originating from reality and going back to reality. There are various descriptions of reality such as the *ultimate reality, ultimate truth* and the *real being*.

There are many controversies about the structure and function of human mind; it has been discussed thoroughly in the previous book *Life and Mind*. We can hold now that the cognitive function of human mind has two main levels called *intellectual mind* and *mystical mind*. Then we would divide intellectual mind into lower part called *philosophic mind* and upper part called *scientific mind*. It has been mentioned earlier that scientific mind characteristically aims to combine abstract ideas with sensory experiences about diverse phenomena in an objective and logical way. This method is used in the cognitive function of philosophic mind also. Hence it is appropriate to say that the deliberation under philosophy is a *superscientific* way of thinking. In contrast, there are two divisions for mystical mind which uses metaphoric, imaginative and emotional way of thinking; these divisions are called *religious mind* and *artistic mind*. [# 1][*].

It may be reiterated that our first-order knowledge about phenomenal universe can be divided mainly into three macro fields; namely, science, religion and art. These are formed by observing the various inanimate objects and living beings through separate faculties of our cognitive mind. Alternatively, the diverse kinds of experiences of phenomena would result in the separate kinds of knowledge. From this point of view, **philosophy is defined as the second-order knowledge which aims to analyze and unify, through *superscientific* manner, the different kinds of first-order knowledge.**

The pursuit of philosophy is primarily meant to explain the existence, development and evolution of the entire spectrum of phenomenal objects, so as to get a vision about the ultimate reality. Now it is expedient to present the **core functions** of philosophy as following.[# 2] [*]

   a) To identify the separate theories which underline the fields of knowledge pertaining to science, religion and art.

b) To analyze and explain the abstract concepts included in such theories so as to compare between the various subjects of knowledge.
c) To develop the method for unifying science, religion and art leading to an integrative view of reality. The final aim of philosophy is to achieve the comprehensive knowledge about the immutable principles of reality underlying the phenomenal worlds.

The word philosophy comes from the two Greek words *philein*, meaning "to love", and *sophia* which means "wisdom". Hence, traditionally, philosophy means "love of wisdom". This definition belongs to the idealist philosophers of ancient Greece, the cradle of western philosophy. According to Socrates, wisdom is the set of noble values acquired from various kinds of knowledge. But, for a person who is dominated by practical and selfish goals of life, it is not easy to understand the terms like *love* and *wisdom*.[# 3]. In this stage of beginning our journey, we may state that the path of philosophy is unclear and tedious due to the following reasons.

Firstly, the philosophers may not agree on the definitions and meanings of the theoretical concepts employed in their fields of enquiry. The social, historical and psychological factors influence human thought especially in the case of abstract ideas. We can easily observe that people have different opinions about the conceptual terms like good, bad, purpose, God, matter and energy. The difference in the elucidations of fundamental concepts will obviously lead to conflicting arguments in the field of philosophy. This is the reason for the various *isms* of philosophy, which stand in the way of producing unified view about phenomenal knowledge.

Secondly, we must consider the tremendous influence of religious faith upon the history of philosophical thought. The approach of religion with regard to reality is to define God as the ultimate being responsible for the creation and working of phenomenal universe. But such descriptions about God come under the field of theology, which has abundantly used many mystical concepts going beyond the ordinary and scientific use of language. We can add that theology includes metaphysical and supernatural words like soul, heaven, hell and sin. When theology was infused into the realm of philosophy it resulted in the crippling of rational thought in philosophy. [# 4].

Thirdly, there are many theologians who propagate the view that theological interpretations or sermons are essential forms of philosophy. This is the reason why the distinctive aims of philosophy as outlined earlier are ignored or subverted. Moreover, the different theologies pertaining to various religions and their subdivisions are mainly responsible for the emergence of conflicting theories about reality.

Fourthly, at other times people use the word 'philosophy' to mean a slogan or maxim that people hold; examples are "Eat, drink, and be merry" and 'Live and let live". It is necessary to be clear about such ordinary usages before we embark on the serious study of philosophy.

Consequent to the above facts, majority of people think naively that philosophy consists of ambiguous ideas which do not serve any practical purpose. So far philosophy has not succeeded to produce definitive notions about the fundamental aspects of nature. It is often alleged that the work of philosophy is to raise doubtful questions as well as to argue at length without reaching a conclusion. This amounts to the popular view that there is no progress in the philosophical knowledge. Though countless number of branches has emerged in this field, during the last three millennia, these thoughts are hovering around the age old questions. In this period, where science and religion dominate the affairs of social life in separate ways, the role of philosophy is often suppressed. The departments of philosophy in universities have ceased to attract bright and inquisitive students.

Our endeavor is to give a reply to the doubting people who ask: what is the use of philosophy? There are many dilemmas in our social life which call for rational assessment and solution. Why do superstitious beliefs and practices prevail in our society that enjoys the benefits of modern science? What is the role of ethics in the fields of human activity? **A mind that is objective and secular only can resort to the way of philosophy in order to solve our social dilemmas.** There is a need of proper criteria for settling the normative issues where divergent arguments are possible due to the vested interests of conflicting groups. The purpose of this book is to rediscover the theoretical framework of philosophy in order to make it practical and efficient for solving the problems of our life.

We have to consider philosophy as an objective and integrative knowledge for reconciling the diverse paths of science, religion and art. In the view, the subject of philosophy can be broadly divided into two branches, namely, *ontology* and *epistemology (theory of knowledge)*. This bifurcation of philosophical thought definitely provides the theoretical framework and method of thought appropriate for the goals of the subject. The progress of philosophy vitally depends upon developing ontology and epistemology in systematic manner. Note that the present chapter focuses on the basic functions of pure philosophy. Hence the various branches of applied philosophy such as political philosophy, environmental philosophy, economic philosophy and feminist philosophy are excluded from our purview. [#5].

2.2 Ontology

The literal meaning of 'ontology' is *science of being*, based on the Greek roots. In simple way the aim of ontology is to propose the theory of reality for integrating the diverse forms of phenomena. For that purpose philosopher has to analyse the structure of things in order to find the fundamental constituents that may be treated as the stuff of universe. We prefer to view the elementary stuff as the form of final unity of diverse phenomena – it is the **reality**. Since phenomena are always changing, they do not have real existence. Hence traditionally reality is construed as the answer to the problem of 'what is', because the ultimate being means 'all that is'. [# 6].

Adopting *scientific* approach we can divide the phenomena of universe into inanimate things and living beings. In the area of science it is the accepted practice to treat the inanimate things as physical objects. As such the *physical* features like length, volume, weight, force and so on are measured through mathematical techniques and these ideas are used to construct the knowledge of science. The basic theory of science is that the physical world is made of matter and energy.

In the next stage, science recognizes the distinctive features of living beings, mainly, plants and animals. Certain activities like metabolism, self maintenance, reproduction and interaction are treated as the characteristic aspects of *life*. In the higher level of animals, especially in humans, the organ called *nervous system* is developed. Brain is the centre of nervous system which produces the mental states like feelings, emotions and ideas. The totality of mental states produced in brain and nerves is called *mind*. It is expedient to hold that life and mind have certain common properties, such as purpose, creativity and freedom, which are nonphysical. We can introduce the term ***consciousness*** to refer to the entity having these nonphysical properties. The term *consciousness* is appropriate for unifying the two levels of phenomena - life and mind - which emerged in the evolutionary history of world.

Consciousness has certain nonphysical aspects like creativity, purpose and freedom which lie outside the scope of science. Being a practical field of study, science tries to explain the levels of consciousness - that is, life and mind - in terms of the physical activities of body. However, the super-scientific or philosophical point of view enables us to distinguish between body and consciousness as the principal parts of a living being. If we scientifically consider the inanimate things and living beings together, the universe has two distinct parts, namely, physical body and nonphysical consciousness.

The fundamental constituents of physical body are matter and energy. But it has been discovered in last century that matter and energy are inter convertible. Hence, we can use the word 'matter' to include energy also, leading to the statement that body is made of matter. Further, according to the traditional thinking in science, **matter** is constituted by tiny particles called atoms.

It is imperative to begin the philosophical subject called ontology by accepting the fact that the stuff of our universe has dual parts – **matter and consciousness.** The inference about the fundamental constituents of universe is based on the observation of phenomenal world. Is it proper to treat matter and consciousness as the dual forms of reality?

In the early stage of philosophical thought, philosophers found that the existence of two opposite realities is very problematic due to the following reasons. If the two realities – matter and consciousness – exist eternally and independently, then there cannot be any connection between them. This negates our daily observation that these opposite forces of reality cooperate and interact to produce the worldly phenomena. Then there must be a third reality for connecting these opposites; in that situation matter and consciousness do not appear to be realities in the first place. Consequently, the idea of two independent realities cannot be sustained.

In order to overcome the above problem of two realities, *the ancient philosophers adopted the idea that one reality is superior to the other.* Then they deliberated upon the question: which of the two fundamental constituents is more real? This led to the view that one of the fundamental constituents exclusively is the reality and that the other constituent has only dependent existence. Accordingly, exponents of ontology were divided into two groups – first group advocated that consciousness is reality while the second group upheld matter as reality. In this manner the rival groups adopt the view of **monism**, because it holds that only one of the fundamental opposites has real existence. The consequent set of ontological doctrines as a whole can be described as **monistic philosophy**. This nomenclature is an innovative idea of this book for distinguishing the erstwhile philosophy against the newly proposed System Philosophy. It is intriguing that, as we will explain in due course, our reference books paradoxically adopt the various notions like monism, dualism and pluralism in order to describe reality. [# 7][*].

Proper elaboration of reality is postponed to chapters 3 and 5. It is now obvious that the various kinds of knowledge are based on separate notions of reality. The reason: *theory is the expression of the particular ontological view adopted.* Accordingly, there is an intimate relation between different notions of reality and the diverse forms of our knowledge about phenomenal things. This relation will be clarified in the following section about epistemology. We would enquire how each of the kingdoms of knowledge gets divided into sub regions (like countries) in tune with the various conceptions of ontology. For that purpose, the relation between ontological divisions and different kinds of knowledge must be analyzed in a systematic manner. It is the function of second branch of philosophy (theory of knowledge / epistemology) to compare and unify the diverse forms of knowledge. This will lead to our innovative idea of worldviews, to be described in due course.

2.3 Theory of Knowledge (Epistemology)

It is expedient to distinguish epistemology from ontology by using a simple example. When I say that this room contains four objects namely chair, table, paper and computer, it is a proposition coming under ontology. The reason is that the room and the objects are parts of this world and hence, pertain to the notion of existence. On the other hand we describe the room as well as the included objects in terms of the various properties experienced by us. The knowledge about room and other objects consist of the descriptions of the concerned properties. *The key point here is that the knowledge about particular objects is essentially different from their existence. We have knowledge about unicorn or angel in terms of a set of definitions of properties. But such knowledge does not imply the existence of unicorn or angel.* Considering the various ways of experiencing things, the same set of things can give rise to different kinds of knowledge belonging to science, religion and art. Additionally, we try to unify these shades of knowledge using the method of philosophy. The analysis, comparison and critical evaluation of various kinds of knowledge would come under epistemology.

Suppose that a person is sitting in a room having four windows with different colours of glass namely green, red, yellow and plain. The person can look through these windows and see the road outside the room in the corresponding colours. Obviously, he sees the real colour of the road, which is black, by looking through the fourth window with plain glass. These different sights of road are similar to the contrasting kinds of knowledge obtained in science, religion, art and philosophy. It is the specific function of epistemology, the second main branch of philosophy, to compare and evaluate between different forms of knowledge about phenomenal objects. Finally, the reality or true aspects of worldly phenomena is revealed through the unification of multiple streams of knowledge. In this integrative exercise, the particular aim of epistemology amounts to determining the validity of different types of propositions. Considering the example of road again, we can ask: is there any validity or factual truth in seeing the road variously as green, red and yellow? It implies that the validity and truth of various kinds of knowledge is a subject requiring deeper levels of epistemological analysis.

We may reiterate that the focus of epistemology is on the knowledge having an organic structure combining the stages of theory (Ty), hypothesis (H), deduction (D), testing (T) and inductive inference (I). This point is intended to highlight the lacuna of traditional approach of epistemology, which has remained in the level of individual propositions.

Initially, different kinds of unitary propositions are formed according to the objectives of understanding. Then the process of specialization and combination is repeated so as to form the larger fields of knowledge. In this context, the specific aim of epistemology or theory of knowledge is to unify the propositions belonging to diverse fields as science, religion and art. For unifying two different objects called table and chair, we must use a higher principle namely 'furniture'. Similarly, epistemology utilizes certain concepts about structure and meaning for comparing the various propositions with regard to different fields of knowledge. These unifying concepts are introduced here in the form of four parts of theory of knowledge – *methodology, source of knowledge, justification and truth.* [# 8][*].

i) **Methodology**

The purpose of methodology is to analyze the organic structure of ideas contained in the five stages namely theory, hypothesis, deduction, testing and inductive inference. It may be emphasized here that the grammar and semantics of language is excluded from the premise of methodology. We know that grammar deals with the categorization and arrangement of words included in a sentence, whereas, semantics analyses the order of meanings. In contrast, methodology is concerned with the structure of ideas expressed in the propositions and its relation to objective world. The above said five stages consist of separate types of propositions, which can be grouped into two sets as under:

- Deductive propositions (DP) including the propositions of Ty, H and D.
- Inductive propositions (IP) including the propositions of T and I.

In a more popular manner, we can say that methodology deals with two ways of understanding things, namely *deduction* and *induction*. In deduction, we start with certain abstract ideas in the form of definitions, mathematics and logic and then deduce a particular principle, like theorem or law. On the other hand, induction is the method of collection and analysis of data pertaining to particular objects, so as to arrive at the inference of general principles.

Primarily, the deductive propositions consist of abstract ideas obtained through rational thinking. The major classifications of such propositions are given below.

- *Axioms (self evident ideas)* – main set of this class are nouns, other words and their meanings as appearing in the dictionary. For example, the word table and its meaning is an abstract conception of our mind.
- *The propositions of mathematics and logic* – the statement 2+4 = 6 and the theorems of geometry pertaining to triangle are cited as examples in this context.
- *The abstract concepts about the characteristics of the universe* – the important instances of this class are the concepts like God, soul, creation, sin, matter, weight, mass and cause-effect. Such

concepts are used to construct the theories of religion, science and art meant for understanding the phenomenal aspects of world as well as its reality.

Using one or more abstract ideas we can construct deductive propositions. Additionally, there is a special method called *syllogism* by which we combine a deductive proposition with certain facts in order to produce a conclusion. This is a creative function of generating new deductive propositions as knowledge. The following example illustrates the method of syllogism.

A All men are mortal *(premise)*
B John is a man *(fact)*
C Therefore, John will die *(conclusion)*

The conclusion that *John will die* is new information derived through rational thinking on the basis of premise and fact. This syllogism is valid if the premise and fact have consistency in meaning. It is clear that the premise (A) is a general principle obtained from our past knowledge. The fact (B) has a special relation to the premise so that the conclusion (C) follows in a rational way. An important feature of the syllogism is that the truth of conclusion is contained in the premise. In other words, the conclusion is necessarily true due to the rational process irrespective of whether the premise is true or false actually. We will explain this point further in a separate section when we take up the aspect of truth. The method of syllogism can be applied in various situations provided the premise is an abstract idea or deductive propositions. It allows us to construct syllogisms about God, soul, angel, heaven, hell and other supernatural entities also. It is remarkable that the conclusion of a syllogism is a kind of new knowledge obtained through the combination of the information of premise and fact.

Now we will consider the **inductive propositions**; these are formed by the fusion of sensory experiences with rational thinking. How do we understand that the length of this pen is four inches? As the first stage, the light rays from the pen reach the eyes which subsequently produce neuron networks in certain part of the brain. Along with such neuron networks, which are called sensory impressions, the ideas about the length of the pen are produced. It involves an intricate process of retrieving from memory the abstract ideas such as pen, length, inch and four as well as the rules of language. Then such ideas are combined with the sense data to produce the inductive proposition about the length of the pen. It is instructive to say that inductive proposition is a generalization of the data collected through sensory experiences. Even in the direct perception of a thing, say pen, there is a stream of data coming into the brain, which undergo the process of generalization.

Next point to be emphasized here is that there are different kinds of combinations of deductive and inductive propositions under science, religion and art. This difference is caused by the separate processes of forming ideas – it can be compared to the different technologies adopted in various departments of a factory. The same set of rays from an object can produce different kinds of sensory experiences and ideas in the respective faculties of science, religion and art. The mental faculty pertaining to science processes the sensory data in an objective manner without involving imagination. On the other hand, the faculties of religion and art normally add the colours of imagination, metaphor and symbolism to the raw data so as to produce peculiar forms of sense experiences. We can add that the religious and artistic experiences are mystical which are in sharp contrast to scientific experiences.

The salient features of deductive and inductive propositions pertaining to different subjects like science, religion, art and philosophy will be studied separately in due course. However, it can be stated now that the enquiry of methodology under content view is divided into two conflicting doctrines, namely, rationalism and empiricism. On the other hand, mysticism and logical positivism are the methodologies under process view. The details of these methodological doctrines will be explained later.

ii) Source of Knowledge

The next part of theory of knowledge is the philosophy of mind with regard to the source of the concerned branch of knowledge. As mentioned earlier, we can expect that the different fields of knowledge such as science, religion, art and philosophy are produced in separate faculties of our mind which are conveniently called scientific mind, religious mind, artistic mind and philosophic mind. The main issue is to describe how a particular faculty of mind exists in order to produce the corresponding type of knowledge. So there is close connection between views on methodology and source. The concerned views of important philosophers will be introduced at appropriate occasions.

iii) Justification

A proposition is a statement about a thing or event belonging to the universe; it is produced by combining sensory experience and rational thinking in various proportions. In order that such a proposition is valid, the corresponding thing or event must exist as a part of universe. The philosophical enquiry about the existence of such objects of knowledge is called justification. To make this point clear, consider the scientific law that the boiling point of water is 100^0C. This law is justified only if water exists as a physical object made of matter. How can we know that matter exists with a set of physical properties? This is generally the problem of justification with regard to scientific laws. Coming to the field of religion, the various propositions are constructed on the basis of the religious concepts like God, soul, sin, heaven and hell. Hence, we may naively think that the justification of religious knowledge rests on the actual existence of such supernatural beings. But how can we know that such supernatural entities exist?

Since the propositions of **inductive inference** are based on sensory data, we can simply say that justification is done by resorting to evidence. Here evidence means the sensory experience that leads to the given proposition. But there are serious difficulties in choosing the evidence and also in determining its efficiency. Specifically, we must address two sets of issues in this context as following.

- What is the criterion of proper evidence? What is the nature of justification?

- Does justification imply truth? Can we say that a justified statement is always true?

In this context, we have to mention the famous definition that ***knowledge is justified true belief (JTB)***, which is attributed to A. J. Ayer (1910-89). It is in tune with the practice of studying the topic from the level of individual propositions. Discussion of the erstwhile writings about the controversies in this field is beyond the scope of this book. A few relevant points are given below.

With regard to the first issue listed above, we must obtain all kinds of evidences so as to rule out the possibility of the proposition to be false. But it can be explained below that this condition is impossible to happen – evidences are fallible - in actual situations.

Our senses have limitations while producing evidences about perceptual knowledge. As sense experience is variable according to the methods of experiment and observation, we can agree that perceptual knowledge is fallible. In the same manner, any verdict based on circumstantial evidence falls short of certainty. Moreover, there is no way of collecting all evidences; obviously the data about future events is unknown since they are yet to happen. For example, we have not so far obtained any evidence about the existence of living organisms in other planets and also in astronomical bodies outside the solar system. Hence the proposition "life exists only on earth" is at best a *fallibly justified belief.* We just don't know the truth about the matter, even after we have examined all evidences at our disposal.

It is now clear that inductive propositions based on sense experience and perceptions are not absolutely certain, mainly because future is unknown. So we are led to the principle: ***inductive knowledge is a belief that is justified by fallible evidences****. Of course, 'fallible' includes the possibility of both truth and falsehood. Now we can reach at the conclusion that justification is not identical to truth, when we consider inductive propositions. The foregoing analysis points to the modified definition: knowledge is a fallibly justified true belief (FJTB).* It amounts to saying that justification and truth are different properties of inductive propositions.

However, the **deductive propositions** have infallible justification since they are obtained exclusively through logical reasoning without involving sensory evidences; in this situation justification is identical to truth. But deduction cannot produce empirical knowledge about particular things, nor does it give scientific laws.

Now we will take up problems about the criterion of proper evidence as well as the nature of justification, as discussed by many proponents of analytic philosophy in last century. We may mention about the thought experiment presented by Edmund Gettier in 1963 to show that there are some instances of FJTB, which involve luck and hence cannot be treated as knowledge. Let us consider the example: Suppose the clock in my room stopped working at 8.30 pm last night. Without being aware of this fact, I glance at the clock in the morning to find that the time is 8.30. Since the time actually is 8.30 by chance, my belief that the time is 8.30 is true. And my belief is justified since I have sensory evidence by observing the clock at that moment. But the *justified true belief* that time is 8.30 is not knowledge, because seeing the clock a bit later would reveal that the clock is not working. We can remove such counter-situations, if we enlarge the evidences by including other corroborative facts.

The idea exemplified here is that the principle of FJTB is not fool proof one, when we consider individual propositions in analytic tradition. In order to avoid the counter examples – *the Gettier cases* - we must give a modified definition: *An inductive proposition is knowledge if it is a fallibly justified true belief (FJTB), which does not involve luck.*

Even when we adopt the norm of fallible justification, there are two challenging issues that constitute an important dilemma about inductive proposition:

- ❖ We cannot decide *how much evidence is sufficient* to declare that the proposition is fallibly justified. The role of additional facts must be admitted in the process of justification.
- ❖ Since a fallibly justified proposition may be false, there is a distinction between justification and truth. If we say that

justification implies truth, this view is known as *infallibilism*. Only deductive propositions are infallible.
- ❖ There is no criterion to determine the *strength* of available evidence. Some of the evidences are based on direct experience of our sensory organs, while some are indirect. So the problem of justification is deeper in the case of invisible entities such as quarks and other quantum phenomena, as well as the concepts like God and soul.

It may be noticed that the above problems of justification arise when we deal with individual propositions in atomist and isolated manner. In this situation, I propose to consider the organic levels of disciplines described earlier in previous chapter. For the philosophical analysis of knowledge, we must focus on the level of **laws**, which are the inductive inferences in various disciplines; each law is formed by the methodological stages Ty, H, D, T and I. It will be clear that we can overcome the issue of *Gettier cases* by the methodology of TyHDTI scheme.

We can note that such issues would still remain unsolved due to the lack of integrative view on the parts of knowledge. ***However, for clearing the age-old riddles, I have proposed to analyze knowledge at the level of laws produced by the stages of Ty, H, D, T and I as well as at the higher levels of concerned disciplines.*** This innovation will pave the way for a systematic analysis of various kinds of knowledge.

We have already indicated the important role of theory (Ty) in the process of justification of laws or knowledge. The innovative idea is that justification pertains to the **existence** of entities represented by the theory. Accordingly, for example, the proposition "this pen has length of 4 inch" is justified only if pen exists as an actual state of affairs of world; it will take us to the question of existence of matter.

The authors of my books of reference have not properly mentioned that justification pertains to the issue of existence of theoretical entities since they follow the tradition of analytic or linguistic philosophy. It can be anticipated here that justification is the most difficult and problematic area of epistemology. In this connection, we have to consider various kinds of **realism** – metaphysical, mystical and scientific – proposed in the history of philosophy for justifying different kinds of knowledge.

iv) Truth

The final stage in epistemology is to deliberate upon the criteria of truth pertaining to deductive propositions and inductive propositions, together with the higher forms of knowledge obtained by combining deduction and induction. A proposition can be treated as knowledge only if it is true; so we come to the question: What is it for a proposition to be true?

A comprehensive understanding of the nature of our mind as well as the theories of justification is required for constructing the correct theory of truth. In this situation it is necessary to introduce the notion of *worldview* with sufficient details. The following chapter will encounter with six different kinds of worldviews, which account for the partial theories of truth causing divisions in the range of human knowledge. Since the problem of truth is to be solved by considering the entire area of philosophy and knowledge, this important subject is reserved for the stage of final chapter.

NOTES of Chapter 2.

\# 1   Four divisions of human cognitive mind – philosophic mind, scientific mind, religious mind and artistic mind -- is my original idea and it is essential to understand the function of philosophy.

\# 2   List of core functions given here would serve to provide a radically new definition of pure philosophy. Many books under reference give traditional definitions of philosophy; but these are generally confusing; as specific examples, see Anthony Harrison–Barbet (1990) and Brooke Noel Moore, et al (2005).

\# 3   See George Thomas White Patrick (1978), page 13.

4   Tarnas (1991) gives a detailed account of the spiritual and intellectual authority of Catholic Church during medieval period of Europe.

5   Main reference books used for chapters 2, 3 and 4 are: Anthony Harrison– Barbet (1990), Brooke Noel Moore and Kenneth Bruder (2005), Chatterjee (1988), Copleston (1994), Davies (2000), Ewing A. C. (1994), George Thomas White Patrick (1978), Grayling A.C. (Editor) (1995), Grayling A.C. (Editor) (1998), Guyer (2008), Hick (1994), Kant (2003), Lavin.T. Z. (1989), Masih. Y (1995), Macquarrie (1985), Max Charlesworth (2006), Russell (1992), Schilpp (Editor) (1941), Schmidt (1967), Tarnas (1991), Taylor (1994), Thilly (2000), Thomson (1997), Urmson J.O. and Johathan Ree (1989), Washburn (1997).

6   A helpful explanation of the concept of reality is available in George Thomas White Patrick (1978). The concept of reality will be discussed thoroughly in chapter 5.

7   In the present chapter, monism is defined in content view. It defers from the monism of Spinoza, which follows process view, to be defined later in fourth chapter.

8   The scheme of four parts of epistemology is my original idea. It would enable us to analyze knowledge comprehensively.

Chapter 3

Worldviews and Versions of Reality

****

3.1 Six Worldviews in Western Philosophy

3.2 Organic Worldview

3.3 Spiritual Process Worldview

3.4 Mechanistic Worldview

3.5 Physical Process Worldview

3.6 Seven Theories of Reality

Author's main original ideas are marked by [*].

The mark [#] gives the number of note at the end

****

Generally speaking, philosophy is an evolution of human thought from mythological speculations to rational and objective ideas. Going back in the history of philosophy, we can note that various philosophical streams originated in the geographical areas of Greece, India and China during the ancient period, specifically from around sixth century BC. But there was a significant difference between these centers of west and east. In India and China, the religious beliefs dominated the realm of philosophy; hence, there was less rational thought as compared to the elucidation of supernatural and mythological principles. In contrast, the ancient Greece was the arena of secular and intellectual development paving the way for objective philosophy. Hence, the characteristic feature of Greek thought is the ascendency of reason and intellect in the enquiry about the existence of objective world. With the passage of time, such doctrines became the foundation of the building of philosophy in western countries as a whole. Taking these facts into account, ***now we will focus on western philosophy only.*** Here our aim is to present a systematic classification of the various streams of Greek philosophy together with subsequent forms of western traditions.

3.1 Six Worldviews in Western Philosophy

We may begin with indicating the preliminary ideas proposed in the sixth and fifth centuries BC by the earliest philosophers of Greek world. All of them were greatly interested in the ontological problem with regard to the origin of the universe. They deliberated upon the fundamental stuff from which all things occurred. Thales (about BC 640-546), who is regarded as the first philosopher of Greece, advanced the theory that water is the original stuff of the world holding that everything arose out of water. Anaximander contradicted this view holding that the universe is made of an infinite and animate substance. Alternatively, Pythagoras (BC 580-500) presented a rather mystical principle that numbers and numerical relations where the essence of things. According to him, numbers have real existence as beings.

Another important development came from the opposite doctrines of Heraclitus (BC 535-475) and Parmenides (born about 515 BC). Heraclitus concluded that everything is changing. Fire represents change. Hence, fire is metaphorically regarded as the original stuff of universe. He wrote, 'one cannot step twice into the same river', since the river is ever-changing with fresh waters. On the contrary, Parmenides argued that permanence is the fundamental character of reality. He held that reality is a mental being endowed with reason; it must be a permanent and single substance out of which the world is made. This doctrine is the forerunner of *idealism*, which is to be discussed later. The riddle of permanence and change was taken seriously by many contemporary thinkers. Empedocles and Anaxagoras suggested that the world contains an intelligent principle called mind or nous, which gives order and motion to things. Then it was the turn of Leucippus and Democritus to propose that the world is made of tiny particles called atoms having the property of motions. This doctrine is known as *atomism* which became the beginning of *materialism* as a theory of reality.

Around BC 450, a group of professional teachers called Sophists shifted the emphasis away from the study of nature to the analysis of knowledge and social aspects. They resorted to *process view* for teaching various opinions about cultural matters, without going into the rational principles – it amounted to relativism. Then it was the turn of Socrates (BC 469-399) who devoted his life to preach the correct approach of acquiring true knowledge. He was mainly concerned with the fundamental questions of ethics and hence tried to define the virtues such as justice, courage and wisdom. The famous Socratic Method is to start from a general opinion and then modify it by proper examples and reasoning in order to reach satisfactory definitions. Socrates did not leave any written text; we know his teachings through the writings of his disciple Plato (BC 428-348). In fact Plato used Socrates as his mouth piece for elucidating his philosophical doctrine.

We can note that the various fundamental ideas about reality of universe have been laid out in the period of Plato and it serves as the foundation for the edifice of western philosophy. Considering the hitherto history of philosophy, it may noted that even enthusiastic teachers and students face great difficulty due to the lack of common denominator for comparing, evaluating and understanding the diverse streams of philosophical thought. In this predicament, we can introduce here the notion of **worldview** as the suitable framework and guide in order to study the subject systematically.

It may be reiterated that different branches of knowledge are produced as the set of inferences, in accordance with the TyHDTI scheme, by assuming separate theories. There is large number of theories in science such as theories about psychological phenomena, biological bodies, genetics, inanimate substances, atoms and molecules, subatomic particles and forces, standard model and more elementary forms of matter. Similarly, the various theologies of different religions and sects are developed based on the theories – basic beliefs - about God, soul and other metaphysical entities. Additionally, the knowledge under art has host of theories about aesthetic experience. We know that the theories of science, religion and art are based on our experiences of phenomena in different ways. *Such theories can be treated as various phenomenal descriptions of reality of universe. Now we will explain the bridge between reality and the set of theories of phenomenal world using the notion of worldviews.*

When we experience and study phenomenal world, we must take into account the relation between body and mind as appearing in inanimate and living things. The phenomenal descriptions about the existence of body and mind and their mutual relations become the *fundamental ideas* upon which the theories of science, religion and art are to be constructed. A particular conception about reality is philosopher's enterprise. Here we define **worldview** *as the method for classifying the totality of theories in accordance with the perspectives of rational view, empirical view, content view and process view.* Accordingly, a worldview is the set of common ideas found in the broadest family of theories. In other words, it contains the essential ideas of all theories having family resemblance.[# 1] [*].

We arrive at the names of worldviews in two steps: Firstly, there are four worldviews namely organic worldview, mechanistic worldview, spiritual process worldview and physical process worldview. Organic worldview is the set of theories about the knowledge of *value*, while the other three worldviews deal with the knowledge of *fact* under various disciplines of science, religion and art. In the second step, it may be admitted that value can be known only through rational thinking under content view. Physical process worldview is purely empirical. But we find that both mechanistic worldview and spiritual process worldview have dual parts of rational view and empirical view. Thus there are **six worldviews** adopted in the history of human thought, which can be arranged as following Table.

Table 1 : The Classification of Six Worldviews [# 2] [*].

Content view	1. OWV - Organic Worldview
	2. MWV-R - Mechanistic Worldview-Rational
	3. MWV-E - Mechanistic Worldview-Empirical
Process view	4. SPWV-R - Spiritual Process Worldview- Rational
	5. SPWV-E - Spiritual Process Worldview-Empirical
	6. PPWV - Physical Process Worldview

The classification of six worldviews enables us to systematically draw the landscape of numerous theories pertaining to value and fact in diverse fields of knowledge. In the following paragraphs we will introduce the **salient features of the worldviews** with reference to the pioneering philosophers of important doctrines.

3.2 Organic Worldview (OWV)

As one of the most ancient conceptions of nature, the *organic worldview* is primarily concerned with the question: *Is there purpose or design in nature?* There is no doubt about the ability of human beings to work with purpose. Most of our voluntary actions -- for example eating, walking, sleeping, reading and writing -- are directed towards certain ends. Similarly, animals, birds and other creatures also display purpose in their activities like making shelter, selecting food items and escaping from enemies. Considering the entire range of organisms of biological world, we can see a hierarchy of complexity pertaining to physical and mental aspects. We can conceive that nature displays purpose in the historical evolution of living beings.

Generally speaking, inanimate things including machines appear to work blindly and mechanically. Hence we do not normally attribute the element of purpose in the things of inanimate world. But through further deliberation, it can be observed that the physical activities of such inanimate things and also the hierarchy of their complexity are caused by certain purpose existing in nature. Hydrogen and oxygen combine to form water, which is essential for the sustenance of life. The chemical reactions between substances produce compounds having entirely new set of properties. In this situation, it is reasonable to hold that the events of inanimate world are governed by purpose imposed from outside. For example, the purpose of a machine lies in the mind of manufacturer who designed it.

We know that purpose includes the element of value since it has opposite directions – good and bad. Some of our activities display good purpose while others show bad purpose. In this situation, how can we account for the opposite aspects of good and bad observable in the phenomenal world? Challenged by this question, a group of philosophers of ancient Greece mainly Parmenides (born about BC 515), Socrates (BC 469-399), Plato (BC 428-348), Aristotle (BC 384-322), believed that the evil can be overcome by the good aspects of nature. Hence, considering the hierarchical organization of inanimate things and living beings, they proposed that universe, consisting of inanimate things and living organisms, has definitely good purpose.

Consider the body of a living being as example. The various parts of body are integrated harmoniously so as to manifest the various functions of life. It is certain that the various parts of a biological body do not work mechanically because they can adjust and interact with external conditions. In the various activities related to life we can observe the purpose of preserving life by avoiding adverse factors. Additionally, the aspects of mind existing in various forms of living beings have the properties of creativity and freedom also. Considering these points, a living being is treated as an *organism* which has the two essential characteristics – harmonious organization of parts and purpose in activities. Extending this principle to the entire range of things, the concerned philosophers were inclined to believe that there is a good goal or purpose in the existence of the components of nature. The term *teleology* represents the aspect of good purpose or design pertaining to nature.

Based on the above points, the philosophical doctrines focusing on the good purpose of nature are traditionally classified as **organic**

worldview. It applies the notion of organism to universe or nature as a whole. If universe is an organism, what is its realty? Since the phenomenal world has purpose, it is reasonable to assume the existence of an agent or being who owns the ideas of purpose. In other words, purpose implies a mind which has that idea. Hence the proponents of OWV hold that the reality is a supreme mind, which possesses all good ideas used for designing the natural things in a hierarchical order.

Here we have to introduce the term **metaphysics**, which traditionally denotes the rational knowledge about the mental reality. There are various streams of metaphysics as we will see in due course.

***Since purpose is the same as goal or value, organic worldview is concerned with the knowledge about value, not about fact.* The ontological and epistemological doctrines of OWV are generally called idealism. As a theory of reality, idealism holds that mind (consciousness) is the original cause or designer of phenomenal world and it exists as a a supernatural and metaphysical being with good purpose.** However in this situation, as explained earlier, idealism assigns only an inferior status to matter, which is the opposite aspect of mind. The theory of knowledge under idealism is alternatively called *rationalism;* it holds that true knowledge about the values of mental reality and phenomenal world is obtained through the rational ideas occurring in our cognitive mind. Additionally, we can state that the theory of justification adopted in organic worldview is metaphysical realism.

In a general way, we can define **metaphysical realism** as the position that supernatural beings and components of matter as well as the fundamental processes have real existence, if such entities are defined using our abstract and rational concepts. This theory is adopted in order to assert that matter as well as various versions of mind exists as different kinds of reality. It can be expected that there are serious confusions when this line of thought is subjected to critical analysis.

Here we may remark that **philosophy of religion** consists of the ontology and epistemology of the mystical knowledge about reality (called as God) as well as the concerned features of religions as a whole. All religious propositions are *mystical* because they are expressed in symbolic, imaginative and metaphoric language. Such propositions are produced by our mystical mind. In the history of western philosophy, the competing ontological doctrines of organic worldview (OWV) and spiritual process worldview (SPWV) are used to interpret the mystical notions of religion. These topics are usually known as *theology*. The two branches of theologies namely *theism and pantheism* correspond to the said worldviews respectively.

Theism is the religious doctrine or theology about the existence of a personal God with the attributes of perfection and omnipotence. According to the idealist thought, the highest level of reason gives knowledge about reality which consists of a hierarchy of values represented by supernatural beings. But in theist religion, God is conceived as the supreme form of supernatural being. Using the metaphysics of idealism, Plato tried to propose rational justification to the religious belief prevalent in his time. This kind of religion is called as *polytheism* and it abounds in mythological gods described under ancient Greek literature. However, Plato's metaphysical and religious theory of reality paved the way for the religious philosophy about theism pertaining to Judaism and Christianity.

The mingling of Greek idealism with Judaism produced the most ancient and original version of theism. The roots of Christian theism belong to the writings of Aristotle, who envisaged strict monotheism within the premise of the rational principles of idealism. Subsequently, Christian religion emerged with various forms of theologies. By synthesizing such theologies St. Augustine (AD 354–430) laid the foundation of theist doctrine of Roman Catholic Church. The structure of Christian theology as well as the corresponding philosophical deliberation grew over the following centuries. Specifically, St. Anselm (1033–1109) and St. Thomas Aquinas (1224-1274) developed interesting arguments for the existence of God as well as the theology of Catholic Church.

3.3 Spiritual Process Worldview (SPWV)

Let us first clarify the difference between *content view* and *process view* (substance and process) by recalling the earlier introduction of these terms. We can get knowledge about a thing in these two ways. Under content view, we treat the thing as a *substance* or being which exists statically with or without component parts. The terms like matter, electron, table, horse, mind, life and consciousness are defined through content view. The theology about transcendent God – that is, theism – is developed through this approach. On the other hand, process view considers the aspects of change of the thing. The term *process* means *activity*. That is, change is a process or activity caused by a set of circumstances or factors, collectively called *context*. Accordingly, process view about a thing consists of the descriptions of the related context causing change. The thing itself is not considered for study.

We can observe that there are two opposite, but interconnected processes in the phenomenal world; these are identified as *material process* and *mental process*. The mental process is manifested mainly in the aspects of life and mind pertaining to living beings. The properties like creativity, purpose and freedom are nonphysical and hence can be treated as expressions of mental process. On the other hand, the physical process consists of changes in the material states of things, resulting in different formations of matter.

On the basis of the data about the relation between material process and mental process in phenomenal realm, we can infer that the reality is also a process synthesizing those opposite phenomenal processes. It is reasonable to hold that the reality is a union of matter and mind. But there is a controversy as to which is superior. One group of philosophers argued that the mental process is predominant, while the physical process has an inferior role in the creative function of reality. They held that reality is a mental or spiritual process. This way of conceiving reality is here called **spiritual process worldview (SPWV)**. It can be further divided into rational and empirical parts denoted by SPWV-R and SPWV-E respectively. [# 3][*].

The metaphysics about process deals with the underlying forces or factors that cause change in worldly phenomena. There are many ontological doctrines in the historical development of *spiritual process worldview-rational* such as Stoicism, Neo-Platonism, Pantheism, Monadology, Historicism and Panentheism. The most influential philosophers in this broad area are Benedict Spinoza (1632–1677), Gottfried Wilhelm Leibniz (1646-1716), Georg Wilhelm Friedrich Hegel (1770-1831), Alfred North Whitehead (1861-1947) and Teilhard de Chardin (1881-1955). There is a practice to use the term ***process philosophy*** for referring to the group of concerned rational or metaphysical doctrines as a whole.

We can note that pantheism was the dominant process doctrine in medieval era. Coming to the modern period, the best known pantheist is Spinoza. Ontologically, pantheism is the counter part of idealism, which is the key doctrine of organic world view. And, its tenets have special relevance for the development of theology from process perspective; in modern parlance we call it as ***process theology***. This rational and religious thought envisages that mental process-reality is identical with the God having the property of immanence. The difference between the secular ontology and process theology is minor one because God is identified as the process reality of universe. The term *process theology* (or alternatively, *process theism*) has become popular consequent upon the publication of the influential book *Process and Reality* in 1929 by A.N. Whitehead (1861-1947). *Panentheism* is the modification of process theology as suggested by Whitehead's followers, mainly Charles Hartshorne (1897-2000).

For a better perspective, we have to remember that religion -- rituals, beliefs, practices and related abstract ideas -- is the product of our mystic mind. But, when erstwhile philosophers and theologians applied rational and scientific thought to the mystical knowledge under religion, there are two opposite doctrines of theology as *theism* and *process theology*, which deal respectively with the transcendence and immanence of God as reality.

It is interesting to observe that there are some special forms of religion, the so called *mystic religions*, which aim to achieve communion with the process reality (immanent God) through direct experience or insight. The best examples of such mystic traditions are Buddhism and Taoism. Additionally, there are certain mystic cults of larger religions such as Kabbalah within Judaism, Sufism within Islam, schools of Hindu mysticism and monasteries of Christian mysticism. Obviously the theologies of these mystic religions or cults are proposed through *empirical way*, since it interprets the sensory aspects of mystic experience.

Next is the area of **empirical spiritual process reality** *(SPWV-E)*. We can now hold that it is largely associated with religious worship, especially under the so called mystic religions. In this case, we have to consider the subjective experience of reality, expressed through metaphoric language. The term **mystic worship theology** is appropriately used here to denote this class of empirical theologies.

Now we face the problem of classifying the doctrines of epistemology pertaining to the theories about immanent reality or God. As a convention in the philosophical literature, the term **mysticism** is applied to refer to the epistemological ideas of process theology and mystic worship generally. In order to contrast between the rational and empirical methodologies adopted, we may denote the epistemology of process theology as *rational mysticism*, while that of mystic worship is called *empirical mysticism*. These two divisions – rational and empirical – of epistemology are primarily concerned about the nature of mystical experience and knowledge as obtained by deductive and inductive methods respectively.

The justification and truth of religious propositions, especially the existence of immanent God, is an important issue in this context. It is accepted that such propositions, being expressed in metaphoric language, cannot be proved by objective evidences nor can it be justified by rational arguments. This is the reason why mysticism is often considered in a derogatory sense. So there is a tendency to treat the mystic traditions and esoteric knowledge as *irrational*, meaning outside the realm of rational and objective thought. But it may be remarked that such knowledge satisfy our emotional requirements and hence occupy a significant role in the social and personal life. However, we have to deliberate upon the methodology of mystic propositions considering it in the wider context of religious propositions.

Additionally there is an area of process philosophy, which is mainly concerned with the subjective experience of a person influenced by the sociopolitical environment – it is normally expressed in the first person perspective. The main divisions of this area are **phenomenology and existentialism.** The various doctrines under process philosophy and mysticism will be explained briefly later in next chapter. However, in the present stage, it is expedient to separate the secular and religious aspects of Spiritual Process Worldview for understanding the doctrines of important philosophers in this field.

3.4 Mechanistic Worldview – MWV

The historical transition from ancient and medieval periods to the modern era happened in Europe through the epoch of Renaissance in fifteenth century. We may recall that the organic and spiritual worldviews asserted that the phenomenal world or nature is controlled by supernatural beings or forces in a rational, moral and purposeful manner. It means that, out of the two fundamental aspects of nature, mind is given real importance while matter is relegated to level of nonbeing. By focusing on the mental aspects of universe, the ancient worldviews dealt with the abstract ideas about *value*. In contrast, the drastic effect of Renaissance was the shift to **the knowledge of *fact* pertaining to the material aspects of nature**.

Subsequently, the scientific revolution of 16th and 17th centuries caused the *origin of science* as the study of nature in physical terms. Here the term 'physical' refers to certain measurable properties of matter such as weight, mass, force, energy, space, time, length and distance. We can say that the words *physical* and *material* have the same meaning. Accordingly the notion of physical world came into intellectual parlance with the assumption that it is made of matter and energy. The general feature of science is to study the regularities in this nature, which are commonly referred to as natural laws. Scientific pursuit aims to express the natural laws in physical terms using the properties of matter and energy – then they are called *physical laws*.

The pioneering scientists held that the motion of material things of physical world happens as per fixed physical laws; so physical world is like a giant machine in the model of a clock. This resulted in the philosophical approach called **mechanistic worldview** for the study of nature. It may be understood that mechanistic worldview is concerned primarily with the inanimate world covering diverse physical phenomena like heat, atmospheric pressure, gases, waves, and chemical properties of substances together with the aspects of light, electricity and magnetism as well as motion of astronomical bodies. This range of scientific disciplines is known as **classical science**.

We can note that physical science emerged in western countries by the combined effect of the weakening of the influence of Christian religion and also by the growing importance of sensory experiences in the context of studying nature. This quest involves the study of nonphysical aspect of nature – life and mind – by reducing it to the physical activities of matter and energy. Accordingly, biological science, psychology and other behavioral sciences also emerged adopting the physical approach. Hence, for practical purpose, all branches of *classical science* accept the underlying assumption that physical world and spiritual world are distinct and independent fields, involving a certain extent of dualism.

The factual propositions about physical world, under classical science, can be formed through the alternative methods of deduction and induction – so we can find the *rational* and *empirical* branches of mechanistic worldview. The concerned ontological doctrines are deism and materialism respectively.

During scientific revolution, the concept of theist God underwent a transition to a rational doctrine of **Deism**. It upholds the belief that there is a good and wise Supreme Being who created the world, with matter and physical laws, but no longer intervenes in it. Accordingly, deism is a philosophical conception of God; it is a theological doctrine.

As a theory of reality, **materialism** holds that the universe is originally composed of an extended and homogenous substance called matter. Different units of this substance with varying size constitute the elements like water, air, iron and gold. The original form of materialism was proposed by Leucippus and Democritus, two philosophers of ancient Greece. According to them, matter consists of tiny and indivisible particles called atoms with the capacity for motion and change – this theory is specifically known as atomism.

Materialism envisaged that all forms of inanimate substances as well as living organisms were formed by the redistribution and combination of atoms. Here, materialism is a form of *metaphysics* because the existence of matter was conceived by rational ideas. Through abstract reasoning, it holds that matter exists really as the substratum of all extended things. Ancient doctrine of materialism negated idealism since it was particularly averse to assuming any creative or spiritual agency exterior to the world of matter. In the modern era, materialism caused the emergence of the empirical branch of mechanistic worldview. It served as the basis for the empirical framework to study the nature objectively. Here *empirical* means *primarily relying on experiment or sensory experience.*

In the course of development of science, materialism was expanded to the study of living beings. In that context, the term **naturalism** came into more popular use, replacing the word materialism. Accordingly, all aspects of nature, including inanimate as well as living phenomena, are to be explained by resorting to the physical laws pertaining to the movements of matter. This is a mechanistic conception of nature without recognizing any nonphysical or supernatural force. Even living organisms are material bodies working according to the laws of physics and chemistry.

Coming to the epistemology of classical science, there are two opposite doctrines namely **rationalism** and **empiricism**, which correspond respectively to deism and materialism (naturalism).

The essence of rationalism is the view that true scientific knowledge consists of deductive propositions conceived by rational mind, combining abstract concepts and logic. Accordingly, inductive propositions are treated as uncertain knowledge. The key ideas of rationalism were proposed by Descartes (1596-1650) and Immanuel Kant (1724-1804).

On the contrary, empiricism contains the arguments for giving primary importance to inductive propositions. The key argument of empiricism is that the deductive propositions - including definitions, mathematics, logic and other abstract concepts - are also derived from sensory methods. Hence, empiricism holds that all components of scientific knowledge come from experimental data. In this way, it aimed to become the epistemological doctrine of classical science.

The roots of **empiricism** can be traced to the writings of Francis Bacon (1561-1626). He considered only the data and inferences based on experimental methods. But the status of mathematics and other abstract ideas was not adequately explained. It may be stated also that his argument of empiricism is later allied to the theory of knowledge termed **pragmatism** proposed by William James (1842-1910). According to James, the methodology and justification of any piece of knowledge must satisfy the purpose of adaptation and survival of human being. In other words, our propositions have truth if it is in accordance with our practical aims.

In the period of Enlightenment during 1650-1770, empiricism was developed to be a theory of scientific knowledge. John Locke (1632-1704), George Berkeley (1685-1753) and David Hume (1711-1760) were the chief exponents of this doctrine.

The salient features of mechanistic worldview including its important ontological and epistemological doctrines will be discussed in ensuing chapter.

3.5 Physical Process Worldview – PPWV

As explained earlier, the *physical (material) process* consists of changes in the material states of things, resulting in different formations of matter and energy. The group of philosophers, who hold that the material process is predominating in the universe, has advocated a unique worldview. It is designated here as ***physical process worldview;*** this term is explained below briefly. [# 4][*]

When the material process is the first cause of all changes in nature, the activities of all inanimate things and living beings are ultimately produced by the changes happening to matter at the fundamental level. Additionally, the evolution of natural things as well as the occurrence of mental phenomena is explained in physical terms with reference to material process. This worldview is the springboard of the scientific method of studying physical phenomena in a historical perspective.

The essential feature of PPWV is the principle that we cannot observe and analyse the structure of a phenomenon statically because it is ever changing. This point of view reflects the ancient philosophy of Heraclitus. However, we have accepted earlier that content view and process view are complementary approaches to the study of a phenomenon. When we define a phenomenon in order to analyse its structure we are adopting the content view. On the contrary, the process view enables us to study the changes happening to the phenomenon from time to time. Since such changes are attributed to the material aspects, the concerned knowledge is *empirical*, consisting of inductive propositions.

Regarding the method of deriving the propositions under PPWV, we can identify three distinct approaches depending on the nature of phenomenon under study.

> ***The activity of components of the thing*** is observed in the dimension of time. Quantum physics pertaining to the activity of subatomic phenomena is the best example. Here, we focus on the historical changes happening to the components of a thing. As an ordinary example, the different models of a car involving change in parts can be described under process view. Similarly, the changes happening to climate over a period of many years are explained with reference to the activity of components of atmosphere. A branch of philosophy of mind called *computer model functionalism* has been developed through this scheme of thought.

> ***The activity of circumstances (external factors)*** that causes change in the structure of a thing is described. Examples are Darwin's theory of biological evolution and Karl Marx's theory of communism pertaining to economic system. The totality of external factors causing change in the phenomenon is described as **context**. In the case of car, we can study its evolution by describing the context, that is, the socio-cultural factors, which caused the technological development of different models in the history of the vehicle. The streams of thought called *phenomenology* and *existentialism* are to be included here.

> ***The activity of language*** is interpreted by observing the change in the meanings of words under various contexts. A knife has

different meanings such as weapon and kitchen equipment depending on the circumstances of its use. The philosophical study with regard to the variations in meanings to words and sentences is a specialized subject variously called as *linguistic analysis* or *analytical philosophy*.

It is important to mention here that, in the early decades of 20th century, physicists discovered that atoms are divisible and they are constituted by three subatomic particles and four basic forces. Additionally, the subatomic entities have a wonderful property called *particle-wave duality*. Moreover matter and energy are inter-convertible, as per Einstein's theory of relativity. These developments in the area of quantum physics prompted theoretical physicists to abandon mechanistic worldview for studying the characteristics of subatomic particles and forces. Alternatively, they proposed the *physical process worldview* as the new paradigm for physics.

To throw more light on this method, consider the case of a fan that works using electricity. For describing the function of this fan, we can draw the picture of its electric circuit and component parts. This picture is the algorithm (flowchart) of fan, and it holds good for different fans of various shapes. In other words, the algorithm of fan has multiple-realisability and the existence of particular fan is not considered. Similarly **quantum physics**, having two main branches as quantum mechanics and quantum cosmology, adopts machine-algorithm model for describing the activities of subatomic phenomena. Since the absolute existence of particle or energy is not considered, the cause-effect relations pertain exclusively to the description of activities – this marks the basic deviation from the classical mechanics. The methodological framework of quantum physics was subsequently adopted in various branches of biological sciences, psychology as well as social sciences.

In this situation, the philosophy of modern science took a radical turn to the study about the *activity of language*. This specialized subject with regard to the variations in meanings of words and sentences is variously called **analytical philosophy** or **linguistic philosophy**. This branch of philosophy was developed in England and United States and it can be rightly called as the mainstream philosophy of 20th century. Accordingly, the philosophers of analytic tradition developed a new epistemology called **logical positivism** for science which became the revolutionary paradigm of quantum physics. This can be treated as the *first division* of analytical philosophy. We will explain the salient features of logical positivism in the next chapter while dealing with scientific knowledge.

The second division of analytical philosophy is **linguistic analysis** which addresses the key questions of language. How do words and sentences get meaning? What is the relation of words and sentences to the external world? The proponents of this approach upheld the view that the meanings of words are derived from the social and cultural environment. So it is necessary to analyse the language with reference to the context of usage for arriving at the meaning of a proposition. We can note that Bertrand Russell (1872-1970) and Ludwig Wittgenstein (1889-1951) are the chief exponents of linguistic analysis. It may be added that the background of linguistic analysis was the contention that reality cannot be described through process view. The metaphysics about process is full of ambiguities as well as metaphoric usages. In this situation, the concerned philosophers of analytical movement argued that philosophy is concerned not with things, but with the way we talk about things, that is, with language. [# 5].

As the most celebrated philosopher of linguistic analysis, Wittgenstein argued that language is a kind of game like chess. He coined the phrase *language game* to denote a particular pattern of rules for arranging words within a given context. The various branches of science, religion and art can be treated as separate language games. It is emphasized that a language game is able to attribute contextual meaning to the concerned propositions. Wittgenstein adds that a social group has many *forms of life* - ways of doing things or situations of culture - which give rise to different language games For example, the Christian form of life produces a particular version of language game that is Christian theology.

It can be noted that the foregoing emphasis on language was developed along with the doctrines of *phenomenology* and *existentialism*. There is a stress on the priority of changing experiences over fixed abstract principles. This attitude has manifested in the pluralism of radical theologies such as liberation theology, black theology, environmental theology and feminist theology. Also we can consider the New Age spiritual cults, which cropped up in recent decades as a reconstruction of ancient pagan religions. Consequently there is erosion in the influence of institutionalized religions, while the secular approach to spirituality has gained support from diverse sections of people. Religious worship is increasingly perceived as a form of human activity for symbolically interpreting the nature of really in accordance with the prevailing *relativism*.

The analysis of language and culture in contextual manner constitutes the landscape of **post modernism**, which expanded its scope during the recent fifty years. It is very difficult to define postmodernism. But we can simply hold that it is the physical process view mainly about art and literature, with limited influence on science and religion also. As such it is the method of empirical, existential, contextual and linguistic interpretation of various kinds of knowledge. The method of **deconstruction** presented by Jacques Derrida is the highlight of this movement; it is appropriate for anthropological and sociological studies.

3.6 Seven Theories of Reality

We have already defined reality in the beginning of this chapter as *the original cause of all phenomena taken as a whole*. Accordingly, reality is self-caused, permanent and infinite. Considering the finiteness of every phenomenon, it is reasonable to conceive that all phenomena in this universe are ultimately originating from reality and going back to reality. However, in the history of philosophical thought, the description of reality has varied in accordance with the six worldviews considered above. In order to understand this fact, we must see that there is an intimate relation between different notions of reality and the diverse forms of our knowledge about phenomenal things.

Knowledge in a particular discipline is produced by the sequence of theory, hypothesis, deduction, testing and inductive inference (the TyHDTI scheme). Here it is necessary to state the principle: *theory is the basis of our knowledge.* But the numerous theories are finally classified into six worldviews. Recall that a worldview is the broadest set of theories with family resemblance. In this situation, we can say that the justification of a worldview - as well as the host of theories coming under it - depends on a particular theory of reality. Generally speaking, knowledge or theory is justified on the basis of an appropriate conception of ontology. Thus, we can link worldview and theory of reality through the function of justification.

The pioneering philosophers deliberated about the existence of reality, treating it variously as being, substance, person or process. They presumed that reality has existence because it is the original cause of all phenomena as a whole. Using logical reasoning as well as speculation, the concerned philosophers articulated the characteristics of reality so as to explain the origin and development of diverse objects of universe. This kind of philosophical thought is alternatively treated as **metaphysics**. In this connection we must consider the concept of *metaphysical realism*. It is the position that when we know the attributes of reality through abstract or rational thinking, we can assert that such reality has existence. We will explain in due course that this argument raises serious philosophical issues in the wake of different notions of reality.

Now let us reconsider the list of six worldviews in order to find the corresponding theories of reality. In the case of OWV, MWV-R, SPWV-R, SPWV-E and PPWV, there are five theories of reality correspondingly. But in the case of mechanistic worldview-empirical (MWV-E), two different theories are proposed about the origin of physical world. First is the doctrine of *materialism* and the second is *intelligent design argument* (IDA). The following table shows the list of seven theories of reality corresponding to the six worldviews. In due course, we will analyze these ontological theories with the objective of synthesizing them.

Table 2: Seven Theories of Reality [# 6][*].

Six Worldviews	Corresponding Theories of Reality
1. Organic Worldview (OWV)	1. Idealism or Theism
2. Mechanistic Worldview- Rational (MWV-R)	2. Deism
3. Mechanistic Worldview- Empirical (MWV-E)	3. Materialism
	4. Intelligent Design Argument (IDA)
4. Spiritual Process Worldview- Rational (SPWV-R)	5. Rational Spiritual Process
5. Spiritual Process Worldview-Empirical (SPWV-E)	6. Empirical Spiritual Process
6. Physical Process Worldview (PPWV)	7. Materialist Process

Based on the foregoing ideas about six worldviews and seven theories of reality, we can give the **basic classification** of the philosophical doctrines of western thought. It may be added that the field of religion is produced by our mystic mind, which has similar ways of thought as in the case of science. The worldviews have given rise to various religious philosophies – rational theism, deism, empirical theism, pantheism and mysticism -- which are also included in this classification scheme. Accordingly, an innovative **Table 3** showing the classification of western philosophy and illustrious original philosophers is given at the end of this section. This will help to study the subject with the perspective of integration. In next chapter, an essential note about the eastern philosophy also is given for comparison. However, the different philosophical systems of India and China cannot be neatly brought into the framework of Table 3.

Table 3 : Worldviews and Divisions of Western Philosophy [# 7][*]

Worldviews	Ontological Doctrines	Epistemological Doctrines
Organic Worldview (*A)	**Idealism** Plato's Idealism Aristotle's Idealism Theism	**Idealism** Theory of Soul Metaphysical-Realism
Mechanistic Worldview-Rationalism: (*B)	**Deism and Metaphysical Realism**	**Rationalism**
Mechanistic Worldview – Empiricism (*C)	**Naturalism**	**Empiricism** Epiphenomenalism Skepticism
Spiritual Process Worldview (Rational and Empirical) (*D)	Stoicism Neoplatonism Pantheism Monadology Historicism Panentheism	**Mysticism.** Phenomenology Existentialism
Physical Process Worldview (*E)	Scientific Realism	Logical positivism Behaviorism Functionalism Linguistic analysis

Main Philosophers

*A - Socrates (469-399 BC), Plato (428-348 BC), Aristotle (384-322 BC), Augustine (354-430), Thomas Aquinas (1225-1274)

*B - Descartes (1596-1650), Immanuel Kant (1724-1804)

*C - John Locke (1632-1704), George Berkeley (1685-1753), David Hume (1711-1776)

*D - Heraclitus (535-475 BC), Zeno (300 BC), Plotinus (AD 205-270), Spinoza (1632-1677), Leibniz (1646-1716), Hegel (1770-1831), Teilhard Chardin (1881-1955), Whitehead (1861-1947), Nietzsche (1844-1900), Sartre (1905-980), Heidegger (1889-1976)

*E - Bertand Russell (1872-1970), Wittgenstein (1889-1951), Thomas Kuhn (1922-1996), Levinas (1905-1995), Michel Foucault (1926-1984), Jacques Derrida(1930-2004)

NOTES of Chapter 3.

1 Tarnas (1991, in *The Passion of the Western Mind*, describes worldviews from historical perspective. According to him there are three worldviews in western cultural history – *the classical, the medieval* and *the modern*. This classification does not take into account the philosophical conflicts in particular eras. Hence we can say that the author used the term *worldview* without precise philosophical meaning. This problem is removed here by the articulation of the six worldviews accounting for the internal strives in western philosophy.

2 These six worldviews will be arranged in a 2x2 table in chapter 5.

\# 3   The division of spiritual process worldview into rational and empirical parts is my original idea. It will help to unify pantheism and mystic worship later using System Philosophy of Religion.

\# 4   The phrase *physical process worldview* is introduced here in order to bring different branches of 20th century thought under one umbrella.

\# 5   Chatterjee (1988), page 29.

\# 6   The linking of worldviews and theories of reality, as shown in Tables 2 and 3, is the crucial step for deliberating upon the ultimate reality of universe.

\# 7   This innovative table will help us to know the historical development of the diverse philosophic doctrines and also to synthesize them using System Philosophy.

CHAPTER 4

WESTERN PHILOSOPHY - KEY DOCTRINES OF GREAT PHILOSOPHERS

****

4.1. Idealism of Plato

4.2 Idealist Philosophy of Aristotle

4.3 Spiritual Process Philosophy

4.4 Theory of Knowledge – Rational Mysticism

4.5 Theory of Knowledge – Empirical Mysticism

4.6 Deism and Rationalism

4.7 Materialism and Empiricism

4.8 A Note on Eastern Philosophy

Author's main original ideas are marked by [*].

The mark [#] gives the number of note at the end

****

For understanding the foregoing concepts of six worldviews and seven theories of reality, it is necessary to know the basic teachings of the great philosophers of western world. The following sections will present the most important doctrines in a lucid and analytical manner so as to serve as the guide for developing the topics of philosophy of science and reality. Scientific realism and logical positivism, which are the ontological and epistemological doctrines falling under Physical Process Worldview, are excluded from this chapter because it has appeared extensively in various parts of earlier books dealing with philosophy of science.

4.1 Idealism of Plato

We can regard Plato (BC 428-348) as the greatest architect of organic worldview and idealism. Plato expounded his thought in about 35 books in the form of dialogues, the most famous being *Republic*. The text of the dialogues abounds in metaphoric and imaginative phrases; hence the following concise paragraphs are based on the interpretations of Plato's philosophy as available in the reference books used for the present article. Moreover, the topics of his political philosophy and ethics are outside our purview due to space constraints.

Plato's doctrine of idealism is a rational thought developed from the religious faith namely, polytheism. The mythological belief in many gods, who are assumed to control the affairs of natural things, led to the distinction between supernatural world and natural world. *Plato tried to conceive the order of eternal world inhabited by supernatural beings, which are responsible for the hierarchy and purpose of perishable things.* [# 1]. Moreover, he assumed that human person is endowed with an immortal and supernatural soul that has the intelligence to know the cosmic order. The question about the relation between the eternal beings and mortal things – particularly between soul and body – is his central theme. As we will see below, the influence of religion on idealist philosophy is very strong. Taking this fact into account, Max Charlesworth describes idealism as a *philosophical religion* because it is a recasting of mythological religion by using secular reasoning and practical evidences. [# 2].

Plato's ontology starts with the distinction between **class** and **particular**. This world consists of various things or particulars which can be grouped into certain classes. A class is a group of particulars having a set of common properties. A common property belonging to a class is called a ***universal***. Plato uses the word **Form** to refer to an unchanging property or universal of a class. For example, the noun 'horse' represents a class of animals possessing certain given properties. In other words, horse is a common name for the set of particular animals constituting a class. Plato noticed that the particular horses may vary in various properties like height, length, color and weight. But the Forms attributed to horse as a class are fixed and eternal.

Similarly, red is a Form because it is a common property present in many instances or particulars of the color *red*. In this manner, the names such as cat, snake, house, man, water and air stand for classes that are defined on the basis of the respective sets of fixed properties or Forms. Specifically, the class called man represents the particular persons having different names and physical features; but the Form of man is fixed and eternal because it distinguishes human beings from other animals. Note that here man is a generic term including both the genders. If we consider the gender aspect, then man is a class that is to be contrasted from the opposite class called woman.

In this way, Plato proposed that a class is composed of a set of eternal Forms, which exist as the reality of the corresponding particulars (or appearances). *In this way Plato envisaged that there is a real world of Forms and it is the reality of the phenomenal particulars.* In some texts, the word Idea is used instead of Form; but it is better to accept the word Form here for clarity because the term Idea may be confused with the ideas occurring in a person. Form can be treated as the property, universal or design pertaining to a class; the set of Forms is the distinguishing feature of a given class in contrast to other classes.

It may be noted from above that Plato distinguishes between reality and phenomena using the terminology of **class** and **particulars** respectively. A class is defined by one or more common properties (Forms) observable in set of particulars. The Forms pertaining to a class is the reality behind the phenomenal particulars. Following the path of Parmenides and Socrates, Plato argued that reality is eternal, permanent and unchangeable. *His ontology or theory of reality consists* *essentially in the view that the Forms have real existence as objects. The Forms exist as substances or beings belonging to the world of reality.* Plato tried to justify this proposition by resorting to the correspondence theory of truth, holding that the Forms existing in philosopher's mind have correspondence with the beings in real world. It is the essential feature of the *metaphysical realism* adopted in his philosophy. In this way, Plato envisages two worlds – the world of real Forms and the world of phenomenal particulars.

How do phenomenal things originate from the Forms of real world? In order to explain this, Plato conceives that the phenomenal things have various properties subsisting on a raw substance popularly known as matter. Consider a block of marble, using which a sculptor makes a statue. The marble block is a raw material without any form or beauty. The sculptor applies his Forms to work upon the marble so as to give it the shape of a statue. Accordingly, the various particular objects of phenomenal world have the common substratum that is **matter**. We can simply define matter as a formless and space filling mass. Though matter is also eternal, it is assumed to have lower level of existence in accordance with idealism. Plato calls matter as **nonbeing** because it does not contain any Form. It may be reiterated that the fixed Forms pertaining to class are responsible for giving the appearance of particular things.

Considering the analogy of statue again, we can see that the quality of the statue depends on the Forms conceived by the sculptor. Though the Forms are fixed and unchanging, statues can be of varying quality or beauty. Plato expresses this point by treating worldly objects as images of eternal Forms. When fixed Forms of class are impressed upon formless matter, the quality of impressions may vary causing the production of a variety of particulars. A particular object is treated as an imperfect copy of the eternal Form.

Next we will consider the question: how did the hierarchy of phenomenal world occur? Plato's conception about the increasing complexity found in inanimate things and living beings is a special aspect of his theory of Forms. He argues that the hierarchy in phenomenal objects is the result of corresponding order in the real world. The eternal Forms are arranged in the pattern of a pyramid where the highest point is occupied by the *Idea of the Good*. When this order of Forms is impressed upon matter, the phenomenal world came into existence in the increasing order of value. Here, Plato equates goodness with end or purpose, emphasizing the proposal that the hierarchical order of world manifest good purpose or teleology. This is the defining principle which characterizes Plato's ontology called **idealism**. As a clarification, idealism is the theory of reality holding that mind (hierarchy of Forms) has good purpose and it accounts for the reality of world. So Plato envisages that the real world is an ideal world.

It is interesting to note that Plato was influenced by the religious doctrine of polytheism that was prevalent in its time. The theory that eternal Forms exist as substances or beings reflects the religious faith in many gods. However, regarding the popular use of the word God, the all powerful and perfect being, we can observe that Plato displays confusion. In his book, the *Timaeus*, Plato talks about a Demiurge, who is a sort of architect constructing this universe out of matter and Forms. Note that Demiurge exists along with matter; it does not agree with the notion of God adopted in the religions following monotheism. In short, Plato's theory about the hierarchy of eternal Forms, with the Idea of the Good at the top, does not correspond with the religious belief about a supreme personal God.

Criticism on Plato's Theory of Reality [# 3][*]

First. The problem of dualism connected with the assertion about the real existence of matter and Forms produces may dilemmas. Note that the totality of Forms may be termed as mind (consciousness). Is there any evidence in holding that matter and mind exist separately and eternally? The **matter-mind dualism** at the metaphysical level throws up serious challenges when we attempt to explain the structure and activity of phenomenal objects, especially living beings. How do consciousness and physical body interact for producing the behaviors representing life and mind? As a caveat, it may be noted that the word 'mind' is used in both contexts of reality and phenomena and it need not create any confusion to the reader.

Second. Plato's theory of Forms envisages the view of pluralism with regard to reality, in tune with the religious doctrine of polytheism. His attempt to unify the variety of Forms through the notions of hierarchy and teleology is not successful because it raises many philosophical issues. Who is the agent giving purpose or teleology to the Forms? How do Forms get coordinated in order to produce the variety of particulars? These questions are not answered in Plato's idealism.

Third. Plato advocates the static existence of eternal Forms and matter adopting content view of knowledge. It patently fails in explaining the phenomenon of evolution in the case of inanimate things and living beings. In other words, Plato's theory is not applicable to the changing aspects of world, which actually belongs to the realm of process view of knowledge

Fourth. Finally, but significantly, we can note that Plato does not throw any light on the *problem of evil*, the fact that there is persistent occurrence of evil in this world. This is the biggest challenge to the doctrine of idealism. We need a drastic modification to the notion of teleology in order to account for evils.

As per Plato's ***epistemology***, there are four levels of knowledge which corresponds to the classes of objects existing in this universe. The levels of knowledge are imagination, perception, intellect and intuition. Plato suggests that the four levels can be represented by the corresponding segments of a vertical line - this doctrine is famously called the *Divided Line*. The segments of knowledge and the corresponding objects of universe are shown in the following Table. [# 4].

Table 1 : Plato's Divided Line and Theory of Forms

Levels of vertical line	Levels of knowledge (Divided Line)	Objects (Theory of Forms)
IV	Intuition	Higher Forms of reality (The idea of the Good)
III	Intellect	Form of the classes of worldly things
II	Perception	Phenomenal things of world
I	Imagination (Opinion)	**Religious faith** (including idols of worship), shadows, paintings and other works of art

The function of *reason* is to acquire knowledge regarding the abstract concepts about worldly things as well as about reality; it consists of *deductive propositions*. Moreover, Plato regarded *intellect* and *intuition* – levels III and IV -- as the successive levels of reason. The second level that is *perception* consists of the inductive propositions generated through sensory observations. On the other hand, religious worship including the idols of gods is included in the class of *imagination* that is the lowest level of knowledge. The religious imagination or knowledge is traditionally referred by the word *faith*. Treating faith along with various forms of art, Plato degraded the value of religious activities.

According to the idealist thought, the highest level of reason gives knowledge about reality which consists of a hierarchy of values as well as the Idea of the Good. The eternal forms of reality exist as metaphysical beings. Here Plato tried to propose rational justification to the religious belief prevalent in his time. This kind of religion is called as *polytheism* and it abounds in mythological gods described under ancient Greek literature. Here we can note an important point. Though Plato devalued religious faith as fanciful imagination, he wanted to prove through the application of reason that many gods exists really. In this manner, Plato presented a metaphysical and religious theory of reality, which contradicted the opposite doctrine called materialism.

As per the Divided Line, there are two levels of reason called intellect and intuition respectively. Plato argues that we get deductive propositions of science through intellect and it is accepted as true knowledge about worldly things. The level of intuition is the supreme stage of reason producing the philosophical vision of reality. *Taking these two segments together we can note that Plato regarded reason as a scientific faculty; so it can be called as* **scientific reason**. But he treats perception and imagination as the products of senses which lie outside the premise of scientific reason, leading to endless controversies in later period.

Technically speaking, the dichotomy between deductive and inductive propositions arises from the respective notions of scientific reason and sense experience. Within the class of inductive propositions, philosophers are mostly interested in the perception about phenomenal things; it is the empirical knowledge under science. Then imagination (opinion) is neglected from the purview of main stream philosophy that focuses on scientific view on world and reality. Here we may examine critically the theory of Divided Line, particularly the notion of scientific reason, advanced by Plato.

The first level of scientific reason is *intellect* which is the source of deductive propositions of science dealing with finite things of world. The second and higher level is *intuition* and it produces the abstract principles about God or reality with the attributes like infinite and transcendent. According to Plato, soul is a singular and metaphysical being with the power of scientific reason. Then there is a difficulty to explain how soul has two levels namely, intellect and intuition, focusing on phenomena and reality respectively. Can scientific reason raise from the knowledge about finite the things to the vision about infinite reality? Plato's arguments for answering this question affirmatively are very ambiguous and allegorical.

The abstract principle about *God* or *Form of the Good* is mysterious and super intelligible. This God is compared to the sun. We cannot look directly at the sun; similarly, it is not possible to describe the features of God in the scientific language of the intellect. So we need the higher faulty of intuition which understands God in a metaphoric way. This line of arguments is expressed through the famous **allegory of the cave**. Plato implies that intuition is not scientific, but is mystical. But Plato wanted to convey the idealist principle that reason can ascend from the level of intellect to the realm of intuition. He envisaged continuity in the two faculties of abstract knowledge. This explains why Plato regarded intellect and intuition as the successive levels of scientific reason. [# 5].

The key part of Plato's epistemology is his philosophy of mind – the notion about the **source** of knowledge – describing the faculties of soul. According to Plato, the soul is a metaphysical being with the capacity of rational thinking. It somehow interacts with body causing the functions of sensory organs. This view envisages the dualism between soul and body. However, the proposal that the sensory knowledge (induction) is produced by body would lead to serious inconsistencies in the subsequent development of philosophy of mind. Most importantly, the conflict between rationalism and empiricism originates from Plato's bifurcation of mind into soul and sensations.

The theory of justification adopted by Plato is metaphysical realism. It amounts to the position that the abstract ideas about classes - alternatively called universals - have real existence as beings, even though such ideas are conceived by human mind. We noted above that Plato upheld his realism by resorting to the *correspondence theory of truth*. However, it is a moot point whether we can justify deductive propositions about reality by resorting to metaphysical realism. Generally speaking it is the *problem of universals*. We can admit that matter also is a universal, though Plato treated it as nonbeing. The issue of metaphysical realism has been discussed critically after Plato by many illustrious philosophers, most importantly by Immanuel Kant in 18th century; we will trace the concerned arguments in due course.

In spite of the controversies and shortcomings mentioned above, Plato could establish a secular and objective philosophy which became the inspiration for his successors. It is generally said that the future course of western thought is the elaboration, refutation or modification of the themes originally introduced by Plato. Now we will focus on the next outstanding philosopher of Greek world.

4.2 Idealist Philosophy of Aristotle

Further advancement of idealism happened through the writings of Aristotle (BC 384-322), who developed his own thought after being the pupil of Plato for 20 years. Aristotle's own theory of reality is contained in his books *Categories* and *Metaphysics*. He introduced the term **metaphysics** to mean the study the nature of reality whereas physics pertains to the realm of phenomenal things. Recall that Plato's theory of Forms envisaged two separate worlds – first is the world of eternal forms and primordial matter and second is the world of sensible things. The individual things or particulars were treated as mere copies of the real Forms. For Plato, the Forms are substances or beings while matter is nonbeing and particulars are appearances. In this theory, there was a serious failure because Plato could not explain how Forms get impressed upon the primordial matter. His proposal that individual things *participate* in real Forms as well as the notions like hierarchy of Forms, Idea of Good and teleology are found unsatisfactory.

In order to tide over the shortcomings of Plato, Aristotle modified the concept of substance in the following manner. For example, consider the statement, *dog is a good animal*. In retrospect, Plato treated the words dog, good and animal as classes or eternal Forms. But Aristotle holds that 'dog' is the subject and it belongs to the category of substance, an existing being. In contrast, good and animal are predicates. These are two properties which subsist on the substance. *According to this view, substance exists primarily, whereas the predicates have secondary existence. This notion of substance and predicates would lead to the modification of Plato's theory of Forms. Aristotle advocates that a substance exists as a concrete individual thing, which is a combination of matter and form.* The various substances like table, dog, man, son and water exist metaphysically by the combination of primordial matter with form in different proportions. In other words, according to Aristotle, matter and form are the complementary aspects of a substance.

The above notion of matter and form as existing in a thing is used by Aristotle to propose his **theory of four causes**. He famously uses the example of a potter who makes a vessel out of clay. The clay is formless and lumpy which can be equated to primordial matter; it is the first cause (the material cause). The form of vessel as existing in the mind of potter is the second cause (the formal cause). The various tools which gives shape to the vessel is the third cause (the efficient cause). Fourthly, the purpose or intention of potter stands as the final cause. Here, Aristotle provides a metaphysical interpretation that the various things in the universe occur on account of four causes.

Subsequently, Aristotle holds that the development of things in a hierarchy is due to different levels of combinations of matter and form. The growth of a tree from seed can be explained in this manner. Seed is the primary substance which is a combination of matter and form. When this seed becomes tree, it is said that the form of tree is imposed on matter (seed). In this way, matter is a general term denoting the lower substance upon which the form of higher substance is imposed. For example, a table is a higher substance formed by the impression of the form of table upon wood which serves as the lower substance that is matter. Considering the sequences of lower levels and higher levels in opposite directions, Aristotle conceives the existence of primordial matter at the lowest level and pure form at the highest level in the spectrum of substances in the universe. The only substance which exists as pure form is God, and it does not have the material component. At the other extreme, there is primordial matter without any form. This leads to the concept of God as the supreme form which acts on the primordial matter to produce the hierarchy of worldly substances.

For Aristotle, God is the unmoved mover or the eternal final cause of the universe. All the forms of worldly things exist firstly in God. Being the unmoved mover, God can be treated as the first cause of worldly things. This philosophical formulation is called cosmological proof for the existence of God like an architect or designer of universe.

It is obviously the philosopher's idea of God to say that God is an architect who works on the eternally existing matter to produce a hierarchy of things. This view does not agree with the monotheism adopted in the religions like Judaism and Christianity, in which God is a person who created world out of nothing. However, Aristotle emphasizes a teleological view of universe by proposing that the forms of things are arranged in the ascending order of good purpose.

As a summary, we can note that Aristotle's philosophy is a clear improvement of Plato's idealism due to the following aspects: [*].

- *The definition of substance*, which is the subject, as a combination of matter and form (Plato regarded all forms as substances without any distinction between subject and predicate).
- *The theory of four causes* leading to the hierarchy of things according to teleology.
- *The concept of God as the prime mover* or the first cause of universe.

Now it is necessary to mention the **epistemological characteristics** of idealism, as proposed by Aristotle and his predecessor Plato. [*].

There is no doubt that the *methodology* is to formulate deductive propositions about the fundamental components of phenomena and reality. It is the same as rationalism, to be explained in a separate section. Regarding *source* (philosophy of mind), these philosophers adopted the metaphysical theory of soul, in tune with their religious belief. Further, the theory of *justification* is metaphysical realism, since it asserts the real existence of matter, form and substance as explained above.

For supporting the said strategy of justification, Aristotle formulated ***three rules of thought***, which are given as below.

1. Law of Identity   : A is A
2. Law of non-contradiction : A is not both B and not-B
3. Law of excluded middle : A either is B or is not-B
   (a proposition is either true or false)

Aristotle holds that everybody must apply these rules of thought for knowing the individual things separately. We may note that these rules pertain to the content view of knowledge about the natural and supernatural phenomena. It means that we can think about things – particles or metaphysical beings – distinctly as parts of universe. Moreover, we assume in a common sense way that such things have existence. The cause-effect relations are framed under the assumption that different things exist separately. In this way, for example, we hold that the words like paper, pen, cow, cat, soul, heaven, hell and God represent distinct objects. As a corollary, the opposite entities are assumed to exist as separate objects. In this way, day and night (or man and woman) are opposite objects which have separate existence. In passing, we may indicate that the above *rules of thought* as well as metaphysical realism are closely related. Its criticism will be given later in the course of this book. [# 6].

4.3 Spiritual Process Philosophy

Here we consider the ontological doctrines under *rational spiritual process worldview (SPWV-R)*. This branch can be alternatively called as *process philosophy, process theology or rational mysticism* We may recall the introduction given in section 3.3. Spiritual process ontology suggests rationally that the mental process predominates in the affairs of world, placing material process in the secondary level. In spite of the ambiguities of this convention, the concerned philosophers upheld the principle that God is the mental reality under process view. It is expedient to give a survey by picking out the most important doctrines about divine immanence in phenomenal universe. [# 7].

The roots of *spiritual process ontology* originally came from the thoughts of Heraclitus and Sophists. However, the first school was the **Stoicism** founded by Zeno (BC 334-262) as an attempt to modify Plato's theory of Forms. In the stoic view, the reality is an intelligent divine force named as Logos which is a unity of matter and form. It represents the universal reason that ordered all things in the world.

Since all human beings shared the divine logos, they are members of a universal community. The stoics formulated a metaphysical doctrine of ethics holding that good actions are those following the natural law that follow the logic of universe. The objective of life is to achieve happiness and tranquility by acting for the benefit of community in accordance with the natural order of things.

Next we may consider **Neo-Platonism** which is an attempt to combine Plato's philosophy with the mystery religions, which were based on the mythologies of ancient Greek-Roman world. Plotinus (about 205-270 AD) is the greatest exponent of this pantheist doctrine. It is based on the mystic notion of The One which is the divine power causing the phenomenal world. Plotinus says that the world is the result of the *emanation* from the One. Hence, the divine power or reality is immanent in the cosmos, producing the hierarchical series of things. We can note that Plotinus translated the Idea of Good, belonging to Plato's philosophy, into the notion of the One and treated it as immanent God in accordance with the spiritual worldview. According to Plotinus, the One is not a personal God, rather it is an infinite power which can be apprehended through mystical experience. It is clear from above that the most important theme of Neo-Platonism is the relation between God and World, which is the traditional problem of one and many. Several interesting versions of this process doctrine occurred in the following periods of history.

An important development in the stream of *spiritual process ontology* is **pantheism** advocated by **Benedict Spinoza** (1632-1677). Using rational arguments, similar to the proof of theorems in geometry, Spinoza tried to counter the body-mind dualism of Descartes and to expound his own doctrine of pantheism. The starting point of his thought is the axiomatic definition of *substance* as that which exists in itself or independently of anything else. So substance is infinite and self-caused. If a thing is finite, then it cannot be called substance as per the definition. In the next step, Spinoza deduced that there is only one substance existing in this universe and it is the substratum of all finite things. This substance can be called God or Nature and it is the source of the appearances of matter and mind in the world. To prove this idea,

Spinoza defines matter and mind as the two *modes* or *modifications* of the self-caused substance. These modes are absolutely independent of each other, but both are manifestations of the same universal reality. In this context, Spinoza's theory is treated as **monism** with regard to the existence of world. It implies that God is the world and world is God.

Since Spinoza used the word *substance* for his process philosophy, it causes great confusion. Recall that substance normally pertains to content view [*]. Here we may interpret that by 'substance' Spinoza means *single process representing reality*. The phenomenal things are emanations from the infinite substance called God. Two opposite processes, called mind and matter, appear in all things of phenomenal world. Everything is the union of mental process and material process; we cannot separate these processes. The hierarchical organization of things is caused by the continuous action of God. So God is the immanent force working in the world. God is not a transcendent being.

The literal meaning of *pantheism* is that everything is God or God is everything. Of course, it is difficult to make sense of this statement since God is a metaphysical process while everything in this world is phenomenal and temporary. Hence the defining aspects of pantheism may be given concisely as following [# 8]:

- God is all things or is in all things.
  That is, all things in their unity constitute God.
- The world is the manifestation of an indwelling spirit, which is God.
- God is present in every part of the natural universe.
  Accordingly, it is said that God is immanent in the universe.

Based on the assertion that God is the reality of nature, it can be construed that everything in nature originates ultimately from God. As such God is present even in the smallest component of nature. In other words, the most elementary part of nature is a manifestation of God. By aggregating at successive levels, the totality of various things of nature can be identified with God.

Though mind and matter are manifestations of the same universal reality, they are absolutely independent processes. Mind does not affect matter and matter does not affect mind – here Spinoza rejects both idealism and materialism. How does Spinoza explain the occurrence of mental states – ideas, emotions, desires and will – in human mind? To answer this question, Spinoza suggests the theory called *psychophysical parallelism* which says that mind and body are two parallel processes of the same eternal substance. When material processes occur in brain, corresponding mental processes also occur as parallel. But there is no interaction between these two processes. In this way, Spinoza refutes the mind-body dualism and *interactionism* advocated by Descartes. According to the mystical metaphysics of Spinoza, God coordinates the parallel processes of body and mind in order to produce the purposeful mental states.

Subsequently, **Gottfried Wilhelm Leibniz** (1646-1716) argued that the foregoing theory about parallel and independent processes is unacceptable. According to Leibniz, there is a harmonious interaction between material process and mental process appearing in everything. He developed a metaphysical theory called *monadology.* It holds that the world is made of monads which are the most elementary constituents of things in the world. Each monad is indivisible, but it is a combination of material process and mental process. Hence, monad can be treated as an organism. The ontology of Leibniz lies in the view that matter and mind do not exist as independent substances but mutually dependent processes constituting a monad. As a clarification, we may note that monad is different from atom which is treated in materialism as the indivisible unit of matter.

The doctrine of monadology proposes that the various things of this universe are formed by the organization of monads in successive stages of complexity. Just like the assembly of a car from the most elementary parts. Since the inanimate things are made of monads, they are assumed to contain the dual processes of matter and mind. This amounts to saying that all things in the hierarchy of inanimate things and living beings have the mental aspects as well as the material aspects. This view is called *panpsychism*; it suggests universal presence of mind or consciousness in all worldly things. The self-conscious human being is the highest in the series of organizations of monads. Leibniz offers rational arguments to hold that God is the supreme monad. But he is ambiguous about the emanation principle that is central to the ontology of spiritual process view. Can we say that God, the supreme monad, is the creator of all other monads in the universe? There are serious inconsistencies in treating God as the supreme monad as well as the creator of world; this issue may not be elaborated here.

Next phase of process philosophy is based on the inward path of truth, the approach to seek reality in the inner world of self, elaborated by the proponents of German Romanticism, mainly the thinkers like Rousseau, Goethe and Schiller. For the Romantic, the world is a unitary organism controlled by the immanent reality, properly termed as Spirit. This world is manifested through the imaginative, emotional, spiritual and cultural experiences of human mind. It is clear that Romanticism is against the mechanistic worldview underlying the scientific knowledge. Johann Gottlieb Fichte (1762-1814) and Friedrich Wilhelm Schelling (1775-1854) proposed the Romantic version of pantheism – it is known as Absolute Idealism or Objective Idealism. Rejecting the view of Spinoza that matter and mind are parallel aspects of the Absolute, Fichte and Schelling envisaged that the mind expresses itself using its opposite (matter) by way of a dialectical process. It is a teleological conception of universe to treat it as the expression of consciousness.

Georg Wilhelm Hegel (1770-1831) further developed Absolute Idealism mainly as a reaction to the rationalism of Immanuel Kant (1724-1804). Here, we have to mention certain points about Kant's philosophy, which is drawn in advance from our forthcoming description of rationalism in section 4.6 below. Kant had argued that our scientific mind can get knowledge only about phenomena; the reality (noumenon / thing-in-itself) cannot be known. Human knowledge is exclusively about the world of experience, not about reality. It implies that metaphysical realism is a wrong position. But the Kantian content view accepted that there is a dualism between reality and phenomenon, and between mind and matter, since it is a static view about the universe under the mechanistic paradigm. In order to overcome those philosophical problems, Hegel followed the path of *spiritual process view* and presented a dynamic notion of reality.[*]

To start with, Hegel criticized Kant's theory that a person cannot know reality using the categories of understanding and hence metaphysics is impossible. According to him, Kant's failure is due to applying the categories of understanding to reality without knowing that the categories are essential aspects of reality. *The consciousness consisting of the categories of understanding cannot know itself. We may clarify this point by stating that a camera cannot take the picture of itself.* Extending this argument, Hegel visualizes that reality is Absolute Mind; it is the absolute thought or infinite consciousness. It is the totality of concepts which manifest in all areas of human experience and knowledge.

Following this way, Hegel provides a systematic description of Absolute Mind or Spirit as the dynamic triad of *thesis, antithesis and synthesis*. Spirit evolves through the triad process creating matter and natural things. Hegel explains that mind and matter (being and nonbeing) are the opposite processes which are in a dialectical relation. Mind is the thesis and matter is the antithesis. That is, matter originates from mind in a historical process. Then there is the synthesis of such opposites; it is manifested as a phenomenal thing, which is qualified as *becoming* also. Thus Hegel expresses the view that everything in this world represents the synthesis or the unity of opposites.

Hegel explains the function of spirit in producing the various worldly things. The spirit or absolute mind acts as thesis which generates its own antithesis (matter). Subsequently, the synthesis is achieved resulting in the production of a thing. This triad process is repeated many times so as to produce the hierarchy of inanimate and living things in the universe. *Everything is the realization of the triad process of evolution.* For example, consider the event of the growth of seed to become tree. Initially, seed exists as the synthesis of the opposite processes of mind and matter. The idea of growth to become tree is the thesis – it is an impulse of seed to attain a higher being. Due to the thesis, the seed moves to its contradiction, that is, not-seed; this is the stage of antithesis. Finally, the stage of synthesis is reached, which is the becoming of tree.

The key point here is that every transformation from one thing to another can be described by the triad. Hegel conceives the triad process in an idealist (teleological) manner, giving predominance to the mental aspect. Moreover it is a secular philosophy about process. Accordingly, Hegel perceives that the mental process (thesis) causes the event of evolution. We may reiterate this ontological principle simply as: *the process of development is in the historical order of thesis-antithesis-synthesis and it produces everything as a synthesis of mind (thesis) and matter (antithesis)*. Hegel's evolutionary metaphysics is especially known as **historicism**, since it tries to explain the changes in every aspect of natural world, including our social institutions, as a historical process of the triad.

In the 20th century, **Alfred North Whitehead** (1861-1947) articulated process theology, effecting certain modifications to Hegel's thought. In his magnum opus, *Process and Reality*, Whitehead called the basic unit of world as *actual entity*. Every actual entity has material and mental aspects; there is no body-mind dualism. According to Whitehead, God itself is an actual entity, which guides the evolution of phenomenal world. The hierarchical order of things is formed by the union of actual entities in successive stages. Even inanimate things like rock, metal and water have material and mental aspects. Hence the idea of physical world cannot arise. It can be added that, without distinguishing between body and mind conceptually, there is no way of separating physical world and God. [# 9].

Teilhard de Chardin (1881-1955), a Jesuit priest under Roman Catholic Church as well as a paleontologist, was the first exponent of process theology for connecting Darwinian Theory of evolution with Christian faith. Teilhard's masterpiece is *The Phenomenon of Man*. It envisages the evolution of nature as an activity of immanent God working on the material world. Evolution produces increasing complexity in worldly things as the combination of spiritual and material aspects. Accordingly, he talks about three spheres of nature developed in chronological order – *geosphere, biosphere* and *noosphere*. This idea is articulated by his principle called the *law of complexity consciousness* which states that an increase in consciousness is always accompanied by an increase in the overall complexity of physical structure. Hence, matter and spirit are intimately connected so as to produce more complicated organisms. [# 10].

The main thrust of Teilhard's process theology is that the novelty and creativity of various stages of biological evolution is the result of God's involvement. This line of thought has been elaborated further by other process theologians like David Ray Griffin and John. F. Haught. They envisage that evolution is the effect of continuous creation by immanent God. We can recall that it is in tune with pantheism, which gives the theology about the dual aspects - *material process* and *mental process* – of phenomenal world in accordance with spiritual process worldview. Hence this theological view vehemently refutes the materialist theory of Darwin that evolution does not display any purpose. The doctrines of Whitehead and Teilhard together can be classified as **evolutionary theology** since it argues that events of cosmological evolution are caused by immanent God.

A more nuanced version of evolutionary theology is called **panentheism**, which can be treated as the process version of deism. Charles Hartshorne, the chief exponent of this doctrine, says that God is both transcendent and immanent. Charles Hartshorne modified the theology of Whitehead, naming his version as panentheism. The meaning of panentheism is 'everything is in God', that is 'world is within God.' In contrast, for pantheism, world is identical to God; for theism, the world is completely external to God. Panentheism is a mediating position between theism and pantheism, holding that the God is an infinite reality of which world is a part. However, it leans towards immanence aspect of God, since the transcendence aspect is not clearly marked. God created this world and subsequently causes its evolutionary development by functioning as an immanent force. This is an attempt to connect the notion of creator God to the evolutionary perspective – this doctrine comes under religious philosophy. Since we are now concerned with the rational and secular aspects of ontology and due to space constraints, the details of religious versions of pantheism are excluded from this chapter.

As per process thought, the finite things of world are formed by elementary events which are themselves the union of material and mental aspects. Leibniz called the elementary event as monad, and Hegel described it as a triad process of thesis–antithesis–synthesis, while Whitehead called it actual entity. Such a unit of world has creativity and freedom, or the power of self-determination. The important point to be noted is that there is no body-mind dualism in the concrete real entity under process view. Even inanimate things like rock, metal and water have material and mental aspects. Hence the idea of physical world cannot arise. So we can add that without distinguishing between body and mind conceptually, there is no way of separating physical world and God. In this situation, it may be recalled that scientific knowledge, as developed in modern period, is anchored on the pillars of materialism and physical process view.

Further we may note that process philosophy suggests the theory of *panpsychism* or *panexperiantialism*. Accordingly every monad or actual entity has got a certain level of consciousness, because it is a combination of mental and material parts. That is, all kinds of natural things, including inanimate things, have mental experience. Whitehead makes use of this principle to interpret parapsychology as well as the religious beliefs about miracles, life after death and immortality of soul. [# 11]

Now it is interesting to note the problem of soul appearing in pantheism. As a major consequence of the process philosophy as well as the doctrine of panpsychism, pantheists deny the existence of immortal soul in a human being. Their position is in tune with the notion of God as a unity of everything. Typically, a pantheist believes in various forms of non-personal immortality, such as 'surviving' in people's memories or becoming part of the continuing cycle of life and death.

The foregoing description of God and soul is at times seen as sufficient reason to treat pantheism as atheistic. At least some theists think so. But we may reiterate that pantheism is a theology or religious doctrine, which is far away from atheism, so as to serve as an alternative to theism. However, theoretically speaking, pantheism does not promote any form of organized religion; there is no prescription about specific rituals of divine worship. The attitude of a pantheist towards God and world is somewhat poetical based on the subjective experience of unity and tranquility. In contrast, the practices of theist religions are public functions under the authority of pastoral teachings and laws.

A general criticism on the foregoing doctrines of *spiritual process philosophy* is required here in order to understand the diverse currents of thought emerged in the modern era. The process view of reality, under rational method of knowledge, suffered from certain drawbacks, which are summarized below. [# 12][*]

- Spinoza, Leibniz, Hegel and related philosophers tried to apply the mathematical way of reasoning to the metaphysical questions. They wrongly wanted to deal with metaphysical process using scientific reason. The concepts like substance, monad, spirit and God are defined as processes in abstract way without any evidence from sensory experiences. The pantheist philosophers asserted that such processes have real existence. *This position is the same as metaphysical realism.* The validity of such realism can be challenged through deeper analysis.
- We normally differentiate between material process and mental process in a living organism. In other words, matter and mind are separate things in the phenomenal world. However, it is a moot point whether matter and mind (or material process and mental process) exists really as separate entities. The assertion of metaphysicians in this regard cannot be supported by empirical evidence.

Obviously, there is an unbridgeable gulf between spiritual philosophy and scientific knowledge. In the prevailing socio-cultural climate of Western Europe, the method of spiritual process was found unsuitable for explaining the physical laws described by science. In this predicament, the doctrines of spiritual worldview were constrained to narrow down its focus to the area of religious philosophy. So they served as rational explanations for religious faith in God, soul and other supernatural entities.

Rational spiritual process is the dominant stream in the religious philosophy of Hinduism. Let us consider the chief doctrine called Vedanta, which holds that reality is Brahman with the attribute of consciousness. However, it is assumed that Brahman contains the power of Maya that accounts for the material aspect of universe. There is an unexplained mystery about the existence of these two opposite forces – Brahman and Maya - as well as about the relation between them. The existence of Brahman cannot be described using the concepts with which we know this World. Hence this form of immanent reality cannot be comprehended objectively. In this context, reality is metaphorically called as *nothingness* or *void* or *sunyata*. We may remember the phrase *neti, neti* (not this, not this) in this regard.

From epistemological perspective, eastern doctrine of *rational spiritual process* teaches that we can know reality through meditation leading to *nirvana*; its meaning is metaphorically given as *knowing the self* by overcoming worldly desires.

4.4 Theory of Knowledge – Rational Mysticism

The aim of epistemology is to consider knowledge expressed as propositions and analyze it with regard to method, source, justification and truth. Having discussed this subject previously in the case of theism, we may expect that pantheism also can be analyzed along similar lines. The methodology of pantheism consists of the formulation of deductive propositions (DP) which can be further classified into theory (Ty), Hypothesis (H), and deduction (D). Here theory means the abstract concepts of process theology as covered earlier.

The purpose of theory is to enable us to produce suitable hypothesis and deductive statements, which together serve to understand the empirical aspects of religion expressed in inductive propositions (IP). There are two stages – Testing (T) and Inference (I) – for deriving inductive propositions, which are empirical beliefs about worship of

God or Nature. However, we are concerned here with the methodology of process theology as a rational doctrine. Recall that we have suggested the phrase **rational mysticism** because it focuses on the analysis of such deductive propositions. [# 13] [*].

Regarding the source or philosophy of mind, the most famous doctrine is the *parallelism* proposed by Spinoza holding that mind and body are two parallel processes of the same eternal reality. It aims to refute the mind-body dualism advocated by Descartes. For Spinoza, mental states like ideas and emotions occur as parallel to corresponding material processes of brain. Accordingly, he refuted the notion of soul or mind as metaphysical being. This view exerted great influence on the later development of psychology about unconscious mind. For explaining religious experiences generally and mystic experience specifically, the function of unconscious mind plays the most fundamental role. However, since this view of mind is associated with *panpsychism,* it has certain drawbacks in accounting for the structure of human mind. Also we can realize that process view of mind is patently incapable of showing the existence of mind as a manufacturing center for various mental states.

Is there justification for the propositions of *spiritual process philosophy*? For linking the justification problem to the existence of immanent God, there are certain special hurdles. We have clarified earlier that *existence* is a static concept, which can be discussed only in content view of scientific thought. When we deal with the process or activity of a thing, the aspect of change is taken into account. It is solely because the thing has static existence for a specific period, we are able to observe it as a distinct object. At the same time the thing undergoes change also. There is nothing which does not change. That is, static existence and dynamic change are complementary aspects of the thing. This rule is applicable in the case of God or reality also – it has the dual aspects of permanence and change. However, the question of existence rightly belongs to the field of content view, specifically theism.

As a sequel to these issues, process philosophers do not have a clear position about the existence of God or reality. The famous authors of process theology, like Leibniz, Whitehead and Hartshorne, have resorted to metaphysical realism regarding the divine process. They do not offer proofs for the reflections about immanent God. But they forcefully assert the conceptual necessity of God as process by suitably modifying the ontological, cosmological and design arguments. In this context, we may mention that Ian Barbour and other followers of Whitehead talked about **critical realism**, which was a new name for the process version of metaphysical realism. Here the term 'critical' stands for the basis of sensory and mystical experience regarding belief in God. This is supported by the assertion that religious belief and scientific knowledge are alternative paradigms for understanding the secrets of nature. Further, realism represents the position that we can know reality through the analysis of and inference from data about experiences. [# 14].

4.5 Theory of Knowledge – Empirical Mysticism

There are some special forms of religion, the so called *mystic religions*, which aim to achieve communion with the process reality (immanent God) through direct experience or insight. The best examples of such mystic traditions are Buddhism and Taoism. Additionally, there are certain mystic cults of larger religions such as Kabbalah within Judaism and Sufism within Islam as well as schools of Hindu mysticism. Obviously the theologies of these mystic religions or cults are proposed through empirical way, since it interprets the sensory aspects of mystic worship.

As a convention in the philosophical literature, the term mysticism is applied to refer to the epistemological ideas of *rational spiritual process* and mystic worship generally. In order to contrast between the rational and empirical methodologies adopted, we may denote the epistemology of rational spiritual process as *rational mysticism*, while that of mystic worship is called *empirical mysticism*. These two divisions – rational and empirical – of epistemology are primarily concerned about the nature of mystical experience and knowledge as obtained by deductive and inductive methods respectively. Here we may adopt the same phrase - empirical mysticism – to denote both epistemology and theory of reality allied with the subjective experience of divinity.

The ancient traditions of spiritual science - such as Ayurveda, astrology, and so on – have been described in the earlier book *Origin of Universe*. We can note here that such subjects assume the working of immanent God in the world observed empirically.

During the second half of 20th century, there is a proliferation of writings about linking spirituality with quantum physics, from process perspective. The pioneer in this field is **David Bohm** (1917-1994) who advanced the theory of *implicate order and hologram* to describe the universal reality. His argument runs concisely as follows:

In the context of quantum mechanics and particle-wave duality, there is connection between the particles at considerable distance. This nonlocal and noncausal relation implies that the subatomic particles are *interconnected wholes*. Accordingly the essential characteristic of universe is *wholeness*, which is revealed in the substructures of matter. In order to conceive the ontological interconnectedness, Bohm develops the notion of *implicate order*, meaning the hidden order or gradations of reality that manifests in the explicate order (the hierarchy of phenomena). The model of the wholeness of reality is the familiar photographic hologram. This implies that everything, including elementary components of matter, has creativity or aspect of consciousness. It is now possible to get knowledge about the material and nonmaterial parts of universe in an integrative manner; some theologians have interpreted this idea as a vital step for unifying science and spirituality. [# 15].

Next we may consider the writings of **Fritjof Capra**. His ideas of connecting matter and mystical reality have been mentioned in the previous books of this series. Now it is pertinent to see the books of **Ervin Laszlo** (1932 -) who gave a new version about the link between spirituality and quantum physics. According to Ervin Laszlo, the subatomic field is similar to the notion of space (akasha) appearing in Indian mysticism. Mystics and sages have long maintained that there exists an interconnecting cosmic field -- *Akashic Field* -- at the level of reality.

In his most important book *Science and the Akashic Field*, published in 2004, Laszlo describes that the recent discoveries about vacuum as quantum phenomenon would provide an empirical view of spiritual process. He proposes that the quantum vacuum is also consciousness. Everything in the universe, even the so called inanimate things, therefore has consciousness. Further, Laszlo says that life happens because it comes from the quantum vacuum. The universe is not a world of separate things and events but is a cosmos that is connected and coherent. This view resembles the earliest spiritual traditions in which the physical world and spiritual experience were both aspects of the same reality. Laszlo reveals the cosmos to be a self-actualizing, self-organizing whole, bringing forth life and consciousness in countless universes. He calls his theory as **systems philosophy;** but we can note that it is an empirical theory under process view. [# 16].

As per the foregoing, we may conclusively hold that there is a sort of realism involved in various forms of mysticism holding that the subjective reality or God exists really. This realism is an anomaly of philosophical thought. If a mystic says that God exists in his private consciousness, it has only metaphorical meaning, which is different from the conventional and scientific notion of existence. In this situation, empirical spiritual process also is not competent to answer the question about the existence of ultimate reality.

Now we may consider the epistemology of the foregoing empirical theology. Here we admit that the religious practices produce mystic experiences of special kind which are known subjectively in first person perspective (FPP). Such experiences represent the awareness about the hidden aspects or secrets of divinity. Special training is required for achieving the mystic experiences. It means that only select groups are initiated to the mystery traditions involving special practices like ascetic life, meditation and other spiritual exercises. In the context of mystic worship, the religious experience has the following features:

1. *Ecstasy or altered state of consciousness*

2. *Religious interpretation as union with God*
3. *The ethical dimension of spiritual transformation*

The experiences like ecstasy and trance are given religious or spiritual meaning, holding that the worshippers achieve union with God or absolute reality. It is regarded as a sort of enlightenment, in FPP, about the secrets of divinity. In Buddhism the practice of meditation seeks to reach *nirvana*, which is interpreted as the insight about the impermanent nature of all things. On the other hand, Advaita Vedanta tradition of Hinduism talks about the union of atman (soul) with Brahman (reality) through mystic practices. In such occasions, the consciousness is fully absorbed into or become identical with God. For the mystic tradition of Taoism, it is the experience of 'nothingness' or 'void' which is the infinite source of supreme creativity. We may add that paradoxical language is often used for expressing the personal experiences of mystic practices, because it is basically non-rational.

The **methodology** of empirical mysticism is to formulate inductive propositions (IP). As per our scheme of epistemology, IP has two stages called Testing (T) and Inference (I), which happen through the mystic worship.

Next is the theory of **source** or philosophy of mind pertaining to the subjective direct experience of divine power. The process view as well as the FPP would not allow the mystic philosopher to articulate a coherent theory about mind or soul. Focusing exclusively on the activity of soul, engaged in mystic worship for its salvation, the mystics were not concerned with the problem of soul's existence. (We may remember that a theory about existence is to be expressed in third-person-perspective or TPP only.) In this situation the mystic traditions are incapable of deliberating upon the static existence of metaphysical entities like God and soul. To cite an example from Eastern mysticism, Sri Buddha taught the doctrines of *momentariness, non-self (anatman)* and *karma*. However it is relevant here to consider a special development in western philosophy.

In order to buttress the idea that super rational and emotional experiences are directed to the awareness about reality of phenomenal world, a new branch of philosophy of mind called **phenomenology** was proposed in the early 20th century. It is a secular discipline concerned with the variety of mental states such as perception, imagination, emotion, desire, willing and thought as experienced privately by a person. Hence phenomenology is the study of mental phenomena in the way they appear in our self-consciousness. Edmund Husserl (1859-1938) is the chief exponent of phenomenology; he tried to describe secularly the various activities of self-consciousness. He asserted that the essential aspect of such subjective experiences is *intentionality*; each mental state is directed towards some object. For example, my desire to have food is directed toward the action of taking food. Building upon the contents of intentionality, Husserl described the various ways of interpretation and meaning of the objects of world, as experienced by self. For that purpose, the aspects of social context, logic and linguistic practice are to be taken into account.

One important problem for phenomenology is that a person does not have first person awareness (he or she does not subjectively feel or experience) about most of the higher order objective mental states like logical and mathematical ideas, habitual actions and the properties of things. When I see that the color of this paper is white, I do not 'feel' my subjective experience of whiteness. The mind abruptly rises from the level of subjective experience to the level of objective proposition in TPP. Hence the phenomenological investigation is not possible in the case of objective mental states. On the other hand, we experience privately the properties – alternatively called qualia in TPP -- of our emotions and feelings. For example, we can distinguish the experiences of fear and pain since they have qualia.

Since our objective knowledge is formed in content view under TPP, we can state that such knowledge is the result of a higher level mental activity as compared to subjective experiences in FPP. *Accordingly phenomenology has limited scope* since it focuses more on the intentionality of emotions, feelings and volition leaving out the higher forms of objective perception and abstract thinking. We can note that phenomenology has lost its importance since its project is merged with the psychological doctrines of functionalism and psychoanalysis. As per current view, intentionality is the essential property of unconscious mind; it is profitably analyzed in TPP covering the entire range of semi-conscious and conscious mental states. In spite of the above said limitations, phenomenology would appear to be the only modern doctrine pertaining to the source of mysticism. The first person account of mystical experiences can be given through the secular method of phenomenology.

Turning to the writings of philosophers succeeding Husserl, we find that the main debate was about the possibility of ontology from the first-person study of phenomenology. When we subjectively examine the feelings and emotions, which occur continuously in a momentary sequence, we are not concerned with the things of natural world. For example, a tree-as-perceived is not identical with the tree existing external to mind. This is the case with the mystic version of God also.

The question whether the irrational subjective experiences are directed to the understanding of God or reality paved the way for a popular philosophy called **existentialism**. It is a philosophical movement that grew and matured in the first half of 20th century with considerable influence on literature, art and religion. It refuted the traditional thought of idealism, materialism, and other views, which deliberated about classes or universal ideas pertaining to things and reality. Such pre-existing philosophies are under TPP and they hold that the *existence* of an entity is its *essence* represented by the properties of the class to which it belongs. In contrast, existentialism is concerned with the separate properties of individual entities as comprehended in the FPP of observer. Consequently it redefines the meaning of existence as the set of properties of the individual entity revealed subjectively in observer's mind. Using the new meaning of existence, the philosophic stand point is that existence must be given priority over essence.

Consider a particular horse. The traditional view is that the horse exists as an individual within the universal class called 'horse' having a set of essential properties like matter, special form of physical body, ability to run fast and so on. The variations in the appearance of separate individual horses are not of interest to the philosopher who adopts

TPP either in the content view or process view. But the proponent of existentialism concentrates on his or her personal subjective experiences about the individual horse. Extending this mode of understanding to the whole of things and events in the world, existentialism asserts that existence of a person (human being) is the totality of subjective experiences. Further the meaning of my life must be conceived from the conscious feelings and emotions occurring to me in FPP, instead of using the objective ideas of religion and ethics.

The main themes of existentialism originated from the writings of Soren Kierkegaard (1813-1855) and Friedrich Nietzsche (1844-1900). Between them there is a sharp division about the concept of God; the former adopted a religious view of reality while the latter was an atheist.

Kierkegaard started by rejecting the historicism of Hegel, which is a form of Pantheism or rational mysticism. He emphasized that every individual has the capacity of will to choose his or her destiny. And, the actions and other events of life cause despair, agony, alienation and anxiety. Human life is not designed for pleasure. The world appears to be irrational and absurd is opposition to Hegel's theory of rationality. As a result, there is a philosophical problem called *existential predicament*. Every individual asks whether there is a way for coming out of the sickening problems of life. Kierkegaard then develops a doctrine of God from FPP.

Nietzsche also scorned Hegel's metaphysics and teleological view of world. He declared that every individual would lose faith in God due to the pressure and depth of despair. Nietzsche expresses this point by the announcement that *God is dead*. The concept of the death of God, Nietzsche adds, would enable us to avoid our child-like dependency upon God. We must now find the courage to become gods in a world without God. According to Nietzsche, the greatest need of present civilization is to develop a new type of individual – the superman -- who will be intellectually and morally strong in order to lift the people from existential problem. As a clean support to atheism, Nietzsche counsels that the superman can overthrow 'the submissive and mediocre "slave" mentality that permeates society and dominates religion'. [# 17]

In the subsequent history of existentialism, the towering philosophers are Martin Heidegger (1889-1976) and Jean Paul Sartre (1905-1980) who respectively follow the religious approach and atheistic path. The explanation of the differences between these thinkers belongs to the discipline namely philosophy of religion.

Certain final remarks about the **justification problem** in empirical mysticism may be given as following. Our specific issue is: can we justify the propositions of mysticism which are in process view under FPP? It is important to recall that the justification of religious proposition genuinely depend on the existence of God. We have already described the issue connected with spiritual process philosophy, where the existence of immanent God cannot be proved through the TPP. Then the situation is worse in the case of mysticism, which consists of inductive proposition under FPP. *The subjective and phenomenological meaning of existence of God does not allow us to conceive the external reality of world.*

To end this discussion on the philosophy of spiritual process view, the most important conclusions may be given: If a mystic says that God exists in his private consciousness, it is not in accordance with the conventional and scientific notion of existence. Since epistemology is a secular discipline in TPP, we can hold that, as in the case of process theology, mystic worship also is not competent to answer the question: Does God exist? There is a sort of *realism* involved in various forms of mystic worship holding that the subjective reality or God exists really from FPP. This realism is an anomaly of philosophical thought. Additionally, existentialism is based on the axiom that the subjective experiences of world are irrational including the feelings of despair, anxiety, etc. It upholds the problem of evil and hence promotes atheism. Now we may turn, in the coming subsections, to the philosophical doctrines about scientific study of physical world.

4.6 Deism and Rationalism

We may reiterate that the organic world view promoted the ontological theory of idealism, which was gradually modified into Theism to become the religious philosophy that dominated the Christian faith of medieval period. During Scientific Revolution, the concept of theist God underwent a transition to a rational doctrine of ***Deism***. It upholds the belief that there is a good and wise Supreme Being who created the world, with matter and physical laws, but no longer intervenes in it. Accordingly, deism is a philosophical conception of God in tune with the spirituality of scientists like Galileo and Newton who wanted to study physical laws through deductive method of mathematics and logic. For facilitating further discussion, it can be concluded that deism has four components:

(i) A super naturalist doctrine of reality – God is a transcendent creator of world with physical laws.
(ii) God, being the creator, can interfere or interrupt the affairs of world through occasional miracles.
(iii) Human being has dual parts, namely soul (mind) and physical body.
(iv) Physical world is a machine consisting of matter particles and physical laws.

There are many contentious issues in the foregoing concept of transcendent God. The knowledge about God is mostly religious and mystical – it cannot be treated as rational like scientific facts. Similarly the belief in miracles, as occasional interruptions of God, also cannot be verified scientifically. These aspects about the nature of God, together with related notions like heaven, sin, soul and life after death, are topics in the philosophy of religion, which is outside the scope of this chapter. In the following paragraphs we would explain briefly the theory of knowledge called Rationalism; it pertains mainly to the rational propositions about God, soul, matter and physical laws.

Descartes

The essence of **rationalism** is the view that true sentences of knowledge are deductive propositions conceived by rational mind combining abstract concepts and logic. Plato, Aristotle and other philosophers of idealism used this methodology for articulating the propositions about reality and world. But in the context of mechanistic worldview, **Descartes** (1596-1650) developed rationalism proposing three basic premises or axioms which are mentioned below.

a) *Cogito ergo sum* (I think, therefore I am). Here 'I' denotes the mind that has the power of reasoning. By the expression 'I am', Descartes assumes that mind exists as a metaphysical being or reality. According to him, mind and soul are the same entity. Doubting and questioning are the basic steps of rational thinking. Descartes accepted this method for developing his philosophy.

b) *God exists really as a being*. God is defined as the creator of physical world and the sum total of all truths. This axiom is based on the doctrine of deism, which holds that God created the physical world with all fixed laws. According to Descartes, the existence of such a God can be conceived by the rational mind. Here he relies on the Christian theology of the scholastic philosophers like Anselm, Thomas Aquinas and others belonging to the period from eleventh to thirteenth centuries.

c) *Matter and physical laws are real, since these are created by God*. To establish this axiom, Descartes suggests that a piece of wax has got primary qualities and secondary qualities. *Primary qualities* are treated as the properties of matter; these are the measurable factors such as height, weight, length and volume. But features like colour, taste and smell pertaining to wax, which are experienced by a person in his mind, are considered as *secondary qualities*. Descartes explains that when wax is melted it loses its secondary qualities, but the matter with primary qualities remains. From the primary qualities such as length and weight, we can deduce the conclusion that matter really exists in wax and it is the creation of God.

From the above axiomatic ideas, we can explain the aspects of methodology, source, justification and truth contained in rationalism as following.

Methodology: The deductive propositions pertaining to theory, hypothesis and deduction constitute true knowledge. According to Descartes, such propositions express our *innate ideas*, which occur in the mind without the help of any sensory experience. Further he argued that physical laws are the cause-effect relations between primary qualities of material things which exist really. On the other hand, inductive propositions – under testing and inductive inference – are uncertain knowledge because it is based on sensory experience of secondary qualities of things.

Source: Descartes holds that the instinctive activities of animals, plants and other lower organisms are not rational, but just a part of life. He is following the teachings of the Catholic Church when he tries to bring emotions, instincts and life under the sphere of mechanical body. As per the axiom *Cogito ergo sum,* human soul is the possessor of innate ideas; it exists independently of mechanical body. Only human being possesses a soul (mind), which is a metaphysical being with the power to think rationally.

Here Descartes advocates his famous doctrine known as **mind-body dualism**. As stated above, the term 'mechanical body' refers to the inanimate things as well as the physical bodies of living beings. There are two parts, human mind (soul) and mechanical body, in the phenomenal world. Cartesian *mind-body dualism* arises by virtue of *metaphysical realism* contained in the position that soul and matter exist as real and separate beings.

Justification and truth: We have already explained that *metaphysical realism* asserts the real existence of things so as to correspond with **universals** (ideas about classes) originating in our rational mind. Its aim is to see God, mind (soul), matter, energy and physical laws as real objects. Hence the *justification of deductive propositions* lies in the view, relying on metaphysical realism, that matter and physical laws are permanent and real. The main conclusion of Descartes' philosophy rooted in realism is that the inanimate world has a structure formed out of matter and physical laws; it is called *body*. The physical world is a huge machine working according to fixed physical laws given by Newtonian mechanics. This world is popularly compared to a clock which is treated as a perfect machine. Regarding truth, rationalists hold that deductive propositions have *necessary truth*; this notion will be explained in due course.

Immanuel Kant

In this context, it may be recalled that the notion about creator God, soul and physical world comes from the religious philosophy called *deism*. It is a theological doctrine involving many logical puzzles. Hence, in later period, the Cartesian rationalism came under severe criticism. Is there is any *justification* for the statements about metaphysical beings assumed in deism? We will examine this question by referring to the critical philosophy of **Immanuel Kant** (1724-1804) who can be rightly regarded as the most important philosopher of modern period. In Kant's book *Critique of Pure Reason* published in 1781, he put forward a philosophy which was revolutionary. Because of this work, which is highly appreciated in Western Philosophy, Kant earned the title as Father of Rationalism. The main aim of *Critique of Pure Reason* was to show that the deductive propositions of science have validity and truth. At the same time, Kant refuted the metaphysical realism of Descartes and others who held that God, soul and matter have real existence. In the following paragraphs we will give a concise account of Kant's arguments, taking firstly his criticism on realism.

For considering the difficult question whether a proposition can represent real existence of things, it is expedient to start with an empirical sentence, for example, "dog is a good animal". Here dog is the **subject** while good and animal are **predicates**, which indicate certain properties or qualities of the subject. In our practical life, we get the idea of the predicate on the basis of our empirical perception of dog. But the question is whether the subject has existence. The dog will have existence only if the predicates are supported by some reliable empirical evidences – such existence is not real, it is phenomenal existence. This is the way of experimental science.

Now suppose we define the words dog, good and animal abstractly by way of deduction and make the deductive proposition, "dog is a good animal". Realists such as Plato, Aristotle and Descartes adopted the outlook of *metaphysical realism*, which is the theory of justification of deductive propositions. When a subject is defined using a set of properties conceived rationally, that subject is assumed to exist really. *It means that existence is also included among the set of properties used to define the subject.* Specifically, Aristotle held that the subject (dog) is a substance, which exists metaphysically as the combination of matter and form – it involves realism. Then the predicates are the properties subsisting on the substance.

As another example for deductive inference, let us think about a flying elephant with wings. Here, the elephant is the subject, while wings and flying are predicates representing certain properties. But the question is whether such an elephant has existence. Basing on the argument that *winged flying elephant* is a substance of matter and form, the realists are prone to hold that *it* has existence.

Here, Kant refutes the realism of earlier philosophers. He asserts: when we define a subject (thing) using a set of predicates, it does not represent existence. That is, existence is not a property included in the set of predicates. Kant expresses this point through the statement: **existence is not a predicate**. When we name an object in our mind, it is the subject of the sum total of certain properties or predicates; it has necessary truth since it is a deductive idea. The subject will have existence only if the predicates are supported by certain empirical evidences. For example, coconut tree is the totality of certain properties. But coconut tree (subject) can either exist or not. Although the subject 'coconut tree' has necessary truth when conceived deductively, it can be said to exist only if there are empirical evidences. In this way Kant points out that predicate and existence are different categories. We generally think that an object really exists if we narrate all the properties of it. This is due to the influence of realism. Kant's greatest philosophical contribution lies in exposing the error of saying that predicate implies existence. [# 18]

The distinction between predicate and existence can be applied to the concept of God. Religious belief holds that the original cause of universe is God, who is a personal being with the attributes of omnipotence and perfection. Here the notion of God is a subject because it is defined in terms of certain attributes or properties. According to Kant, the set of predicates does not include existence. He means that when the original cause is defined as God, a being having certain properties, then it is a phenomenal concept. But the view of religion that God has existence is a form of realism which equates predicate with existence. Now we know that when the original cause is described as a being called God having certain attributes, it is a phenomenal idea occurring in our mind. In this situation, as Kant suggests, the belief that God is real cannot be admitted. *It is clear that the notion of God is the essential part of religious knowledge and it cannot be clubbed with rational ideas about physical world. This point is important when we consider the conflict between religion and science.* [*]

Through his critical philosophy, Immanuel Kant exposed the fallacy of Plato's idealist theory of reality, which says that the real existence of Mind is known through the dialectics of our rational faculty. Plato envisaged that the rational mind can ascent from abstract concepts about world to the idea of mental reality. In an ingenious way Kant proved this ontology to be false. **What is real cannot be described using predicates; the entities for which we can conceive properties or predicates are phenomenal.**

Actually Immanuel Kant became the killer of realism. Now we will consider the key points in Kant's philosophy which strengthens the deductive frame work of mechanistic science. The avowed aim of these arguments is to assert that true knowledge is possible in scientific field due to the following reasons.

- The basic ideas of scientific thought are called *twelve categories of understanding* and they are grouped into four classes, namely cause-effect relation, quantity, quality and modality.

- Our scientific mind has a rational structure constituted by the intuitions about space and time together with twelve categories of understanding.
- The rational structure of scientific mind has creativity and it can be compared to a factory. Sensory data constitute the raw material for this factory which produces the outputs – firstly the deductive propositions (DP) including physical concepts and secondly the inductive propositions (IP). Scientific propositions have rationality because these are the products of rational mind. That is, both DP and IP have logical structure.
- Though scientific propositions are logically sound, these are the constructions of our scientific mind and hence are phenomenal.

The abstract ideas like space, time, motion, weight, mass, matter, speed, density and dimensions are physical properties (predicates) conceived by us in the phenomenal realm. In this way Kant establishes that science can achieve knowledge only about the phenomenal aspects of physical world. Since there is no sense impressions about pure matter (thing-in-itself / noumenon), the real existence of matter cannot be known. Similarly the knowledge about our own mental faculties is also phenomenal. This means that the ideas about mental states and activities are the constructions of self. There is no means to prove that self exists as a metaphysical being. As a corollary, the physical laws are also phenomenal constructions of scientific mind. In this manner Kant asserts that the knowledge about matter, physical laws and mind (self) are phenomenal ideas. Hence, the reality of physical universe cannot be known scientifically. This line of thought advanced by Kant is known as *agnosticism* which is surely the rationalist version of skepticism.

Our Key Points of Criticism are the following:

➤ It is a significant contribution to philosophy of science when Kant showed that scientific propositions (DP and IP) are logically sound. But Kant failed to account for the relative difference between DP and IP. He could not bridge the gap between necessary and contingent truth, pertaining to DP and IP respectively. Thus the conflict between rationalism and empiricism continues.

➤ Rational principles about the fundamental aspects of universe are expressed in terms of God, Soul, matter, energy and so on. Kant established that these aspects are phenomenal objects because they have been defined in terms of certain separate properties conceived by our mind. By resorting to realism, we normally treat that such entities have existence. The reason for leaning on realism is that we follow the *three rules of thought* originally systematized by Aristotle. But the rejection of realism through the philosophy of Kant has far reaching consequences in our enquiry about reality of universe.

➤ The philosophers of later period are troubled by Kant's agnostic philosophy about physical world. They are concerned with the issue of justification of science as we have explained in foregoing books in this series.

4.7 Materialism and Empiricism

We may now start by examining the philosophy of mind known as *epiphenomenalism*, since materialism and naturalism have been introduced in previous chapter. It was Thomas Hobbes (1588–1679) who first treated mental states, like ideas and emotions, as the effect of physical processes of brain and nervous system. Mind is a by-product, or epiphenomenon, of material activities of the components of body, just like the smoke produced when a locomotive works. This theory about the relation between mind and body is called *epiphenomenalism*, according to which all knowledge came from physical processes of mind.

For explaining *epiphenomenalism*, we must first consider the development of classical biology on the basis of materialism and naturalism. Accordingly, the biological activities like metabolism and respiration are sought to be explained by the mechanical and chemical activities happening in various organs. The same approach is extended to the study of mind. Using modern terminology, we can define *epiphenomenalism* as the view that mental states are the byproducts of the chemical processes taking place in neuron cells of nervous system. In other words, when certain neural networks are formed, the mental states like emotions, feelings and ideas occur as epiphenomenon. According to this doctrine, mental states are compared to the smoke coming out of burning wood.

In the historical context, the doctrine of *epiphenomenalism* was instrumental for Wilhelm Wundt (1832-1920) to found the method of psychology when he established in 1879 a laboratory in the German town of Leipzig. Thus psychology originated as an empirical study of mental states.

Here the key point is that, based on the view of materialism, the activities of neural networks in the brain are treated as physical processes. And, the material world is causally self-contained; every material event is caused exclusively by other material events. The existence of any supernatural or nonphysical force is ruled out. Consequently, the so-called mental events are to be treated to be the effects of certain material causes. This kind of materialist theory of mind suggests that we can explain the mental activities by appealing to material processes happening in brain. There are two main purposes behind this approach. Firstly, it tries to tide over the philosophical difficulty pertaining to Cartesian *dualism* and *interactionism*. Secondly, neuroscientists could utilize this theory to strengthen the study of brain through experimental methods. They want to detect the physical processes in brain that would help them to explain the mental states in a practical manner, without involving the nonphysical concept of consciousness.

A few important points of criticism are to be mentioned here. Considering the analogy of burning wood and smoke, we can see a one-way causality from burning wood to smoke; there is no causality in reverse direction. Similarly, the materialist process of brain is said to cause the mental states; but mental states cannot affect the brain processes. Epiphenomenalism considers mind as causally inert. This one-way causality goes against our daily observation that certain mental states like desire and willing are the causes of our actions. When I want to open the door, I get up from my chair and walk up to the door and turn its handle. So the main defect of epiphenomenalism is its inability to account for our various voluntary actions. We can reasonably think that there is a two-way causality between neuron networks and mental states.

Additionally, the reliance on mechanism cannot be justified since we certainly know that our mental states have self-consciousness involving the properties like creativity, purpose and freedom. Since mind or consciousness is essentially a nonphysical phenomenon, it cannot be caused by physical processes. The key point here is that, if we study brain through mechanistic science, then it is impossible to account for the occurrence of nonphysical mental states. Obviously there is an *explanatory gap* between material brain and nonphysical consciousness or between matter and mind.

With these remarks, it is expedient to elucidate the theory of knowledge called **empiricism**, as originally proposed by the philosophers of Enlightenment period.

John Locke vehemently denied the metaphysical theory of innate ideas proposed by Descartes and other rationalists. His fundamental thesis is that **human mind at birth is a *tabula rasa* or blank slate** and that all our ideas come from sensory experience and logical reflection. For Locke, reflection includes memory, calculation and imagination as well as the process of combining ideas. Consider the example of a table in front of a person. When the light rays reach his eye (sense organ), it creates certain impressions in brain and subsequently the idea of table is generated. For distinguishing the table from other objects like chair, pen and room, the activity of reflection is necessary. In other words, simple ideas and complex ideas are produced in mind as the epiphenomenon of the physical process of brain. Here Locke implies that the so called deductive propositions are included in the set of complex ideas derived through inductive method.

By resorting to the distinction between primary and secondary qualities of things, Locke adopts *naïve realism* to hold that matter exists as the substratum of primary qualities like weight, length and size.

Additionally, Locke asserts that human mind can be aware of its own ideas only. In the case of perception of an external object, the concerned ideas are accurate copies or representations of the object. So we perceive objects indirectly through our internal ideas. This theory is often called *representative realism;* it is the empirical version of Cartesian theory about physical world. We will see below that Locke's common sense view of justification of inductive knowledge suffers from serious drawbacks.

The principal aim of **George Berkeley**, while developing his empiricist philosophy, is to refute materialism and atheism. He started with the position of Locke that we can know an external thing only indirectly on the basis of our ideas about its sensible qualities. Then, Berkeley argued, we cannot know that the thing has a substratum called matter, since we do not get any sense data about matter. Obviously Locke's common sense theory would lead to skepticism regarding the reality of physical world. In addition to skepticism, materialism leads to atheism – the belief that God does not exist. If matter existed as an independent substance, it implies that matter exists alongside of God. This limitation of God is a ground for the position of atheism. In order to tide over these problems, Berkeley introduced the classical doctrine called **phenomenalism**, which is formally defined as the view that human knowledge is confined to the appearances (phenomena) presented to the senses.

According to George Berkeley, there is no need of distinguishing between primary and secondary qualities of things, because all qualities are ideas occurring in our mind. Even if we divide a thing repeatedly, we can experience the sensory qualities only. Since all these qualities are ideas originated in the observer's mind, an object is the sum total of its qualities perceived by mind. For example, orange is a thing with particular size, round shape, taste and smell. We are not entitled to say that it is made of a real and unperceived substance called matter.

Berkeley upholds the subjective characteristics of knowledge by arguing that all objects in the physical world are mere ideas in our mind. Every material object is a phenomenon (the totality of qualities experienced). Matter is an abstract idea inferred from sensible qualities. Hence, the question whether matter really exists should be avoided from science. When we say that a thing exists, we mean that such and such qualities of the thing are perceived. Accordingly, Berkeley's phenomenalism is expressed by the principle: **to be is to be perceived** (*esse est percipi*).

As per the argument of Berkeley, when the existence of things is defined in terms of perceived qualities or ideas, it is absurd to talk about an unperceived or metaphysical substance called matter. At the same time, he asserts about the existence of mind as an active substance causing ideas. The external world is known through the subjective ideas formed by mind. What about the existence of things in this world, if there is no human being to observe them? Berkeley answers this question saying that such things have existence because they are perceived by God. His *phenomenalism* seeks to overcome skepticism and atheism by resorting to the reality of God. In this milieu, the philosophy of Berkeley is described as *subjective idealism*.

The forgoing combination of empiricism and idealism was severely criticized later by David Hume and Immanuel Kant; both established that there is no escape from the problem of skepticism as far as mechanistic worldview is concerned. We will now take up the philosophy of **David Hume** who was the most systematic and influential among the empiricists. The main arguments of David Hume based on materialism, naturalism and epiphenomenalism are summarized below.

> a) Ideas are generated as a consequence of the impressions in sensory organs and brain. Firstly, *simple ideas* like heat, cold, length, breadth and sound are formed. When such impressions and ideas occur repeatedly, brain has a habit to make generalizations. It results in the formation of *complex ideas* such as mathematics, logic, scientific definitions, various other deductive inferences and cause-effect relations. In this manner Hume presents his principle called *psychological habit of association of ideas*. He suggests that the so called deductive propositions belong to set of complex ideas, which are ultimately based on the physical processes in brain obeying the Newton's laws of motion. In this way Hume denies the metaphysical notion of soul.

b) The analysis of cause-effect relations given in Hume's books relies on the above said psychological habit. For example, a boy observes that bulb lights after switch is on. If the activity of putting on the switch is repeated many times, the boy gets the habit of associating the act of switch with lighting of bulb. From this habit the notion of cause and effect is formed; it is expressed as "when switch is on (cause), bulb lights (effect)". The reason for treating it as a psychological habit, according to Hume, is that many of the actual causes such as flow of electricity, presence of wire, and quality of bulb are not considered. In this way Hume argues that the scientific notion of cause-effect is a purely empirical and habitual idea related to the temporal relation between events occurring before and after.

c) Scientific laws are beliefs generated in human mind due to the habit of association of ideas. The working of sensory organs and brain according to physical principles is the source for scientific knowledge. In this way Hume argues that human mind can know only the phenomenal aspects of physical world as it affect our sensory organs. Out past experiences of world has prompted us to conceive certain abbreviations. The idea of *matter* is one such abbreviation which represents the repeated observation of extended things. Since the idea of matter is only an abbreviation of such qualities as length, breadth, volume and weight, which we experience in physical things, there is no justification for holding that matter exists really.

Now We May Present the Criticism on the Tenets of Empiricism:

The previous books in this series -- Origin of Universe and Life and Mind – have developed the important critical points about empiricism. However, it is useful to highlight here certain aspects especially relevant to the understanding of scientific phenomena.

First, it is pertinent to explain the issue of conceiving matter as an abbreviation of empirical qualities. Consider the question: why is iron much heavier than the same volume of water? The notions of weight and volume are not enough to answer this question. Here we must introduce a new concept called *mass* denoting the stuff (material content) of a thing. So we can say that iron has more mass as compared to that of the same volume of water. It is important to note that mass is not an abbreviation of such qualities as length, breadth, volume and weight, which we experience in physical things. We can state that mass is not an empirical concept under classical science; rather it is a rational or deductive concept.

Second, let us see the empiricist position that deductive propositions can be derived from inductive ideas or sensory experience. Here we may recall the distinction between deductive propositions (DP) and inductive propositions (IP) as explained earlier. For example, consider the case of mathematics and logic, which are obviously based on deductive method. Such propositions are treated as *a priori*, meaning derived without the help of sensory data. Once we know the rules of arithmetic, we can calculate that 5196 + 85472 = 90668, without actually counting. It is a grave mistake, when empiricists ignored the categorical distinction between deduction and induction.[# 19]

Third, the doctrine of epiphenomenalism has serious defects, as mentioned before, mainly because it does not take into account the nonphysical or creative aspect of process of deduction.

Fourth, David Hume is troubled by an important aspect known as ***the problem of induction***. Inductive inferences are based on limited number of observations of sample. There is a possibility that the generalization from sample to population may not hold good in future. Some future evidence may refute the scientific law leading to its rejection. In this way Hume says that *sun may not rise tomorrow*. Science does not have absolute power or logic for predicting future events because physical laws are inferences based on past observations. This view aggravates the problem of *skepticism*. Science is a framework of physical laws pertaining to phenomenal qualities of world and it is patently incapable of talking about reality. The skeptical philosophy of David Hume exerted devastating blow to the epistemology of science.

Hence we notice that the most important problem of empiricism is about the justification of scientific laws. Prior to 20th century, under classical science, unobservable atoms were treated as the forms of pure matter. But the empirical method of science cannot show that pure matter (atoms) exists as the substratum of all physical properties of things. Though matter is earlier assumed to be a homogenous substance, it is now empirically or scientifically observed that there are different kinds of fundamental particles. About 110 types of atoms have been discovered and these atoms are made of many different kinds of subatomic particles. In this connection, it may be added that materialism or naturalism is a scientific theory which promotes **pluralism**.

This scientific view about multiplicity of forms of matter will lead to serious philosophical dilemmas. Mainly, it leads to the skepticism. In the first phase, the concerned empiricist philosophers adopted the position of **naïve realism** as the theory of justification. This is a practical principle, which holds that the empirical properties indicate the real existence of atoms. It may be recalled that the position of naïve realism was later modified to scien*tific realism in the* wake of modern physics. Further, we have already examined whether the tenets of naturalism are satisfactory in explaining the inanimate, biological and mental phenomena.

In this milieu, the development of classical science through empirical methods removed God from the realm of science. The recent centuries witnessed the growing autonomy and competence of physical science to explain the laws of nature, without invoking the role of divine agency. Drawing energy from the ideologies of materialism and naïve realism, a group of thinkers and ordinary people began to subscribe to the belief in ***atheism,*** which says that God and other metaphysical entities do not exist.

4.8 A Note on Eastern Philosophy

The brief and critical review on the ontological fractions given above would not be complete without considering the rich area of *eastern philosophy*, flourished in India and China of ancient period. The creative period of Indian philosophy was the sixteen centuries from sixth century B.C. to tenth century AD when diverse systems of thought were developed. There was greater attention to the ontological issues in these philosophical systems, though it contains vast discussions on various aspects of epistemology also. [# 20]

In general Indian philosophy has nine streams, which are divided into two groups called 'orthodox' and 'heterodox'. The first group was based on the metaphysical concepts of Hindu religion found in the Holy Scriptures namely the Vedas and Upanishads. There are six systems of philosophy in the orthodox group; these are *Samkhya, Vaisesika, Nyaya, Yoga, Mimamsa and Vedanta*. On the other hand, the three heterodox systems are Buddhism, Jainism and Charvaka (Lokayata) materialism.

As a first approximation, we can say that excluding Charvaka materialism the other eight systems are various kinds of religious philosophies pertaining to the religions called Hinduism, Buddhism and Jainism. The dominant feature of these religious philosophies is the ontology of *pantheism* as well as the epistemology of *mysticism*. But we can see the strains of polytheism also in tune with the practices of religious worship. Moreover, materialist pluralism is intertwined with metaphysics upholding the mystic perspective. **So it is not possible for Indian philosophical systems to get neatly placed in our table of ontological divisions**.

Chinese philosophy is a curious mixture of moral philosophy and mysticism in tune with the ethnic religions of China. It consisted of many ancient schools of thought, deliberating more about the political and ethical aspects of human social life, but less about metaphysics and epistemology. The most important moral doctrine is *Confucianism*, propagated by Confucius (BC 551-479). Contemporaneously, there developed a school called *Taoism*, which advocated a process view of reality. The word *tao* means 'the way' and it signifies the cyclical order of cosmic process. Thirdly, we can consider the *yin-yang school*, which teaches that the cosmic order is caused by the opposites called yin and yang. The female and passive aspect of vital energy is represented by yin, while yang refers to the male and active aspect.

It is a sad fact that philosophical thought failed to make any progress in India and China during the modern period, due to various socio-political reasons. In this situation, the intellectual discourse for linking science with philosophy did not arise in Eastern philosophy.

But we want to treat philosophy as a second-order knowledge meant for unifying science, religion and art. For achieving this objective, it is necessary to synthesize the various doctrines falling under the ontological division of western philosophy.

NOTES of Chapter 4

1   The influence of polytheism on Plato's *theory of forms* has not been properly highlighted in the books under reference.

2   Max Charlesworth (2006), *Philosophy and Religion- From Plato to Postmodernism*, pages 12-20.

3   The points of criticism given here are my original ideas, as compared to the relevant passages in the reference books; for example, Grayling A.C. (Editor) (1995), p. 381-388.

4   Plato mentions about Divided Line in Book VI of the *Republic*. See the description in Lavin (1989), pages 31-42.

5   Max Charlesworth (2006), pages 17 – 19.

6   The close relation between rules of thought and metaphysical realism would be hammered later by Immanuel Kant. We will carry forward his criticism in order to develop the System Philosophy, which avoids realism and accepts the complementary existence of opposites like matter and consciousness.

7   Here the main reference are Macquarrie (1985), Tarnas (1991) and Thilly (2000).

8   Macquarrie (1985), page 51.

9   Whitehead, Alfred North (1978), *Process and Reality*, page 348. Process theology is recently elaborated in Griffin (2000), Haught (2000) and Haught (2001)

10   Teilhard Chardin (1999), *Phenomenon of Man*, (Sussex Academic Press, 1999).

11   Griffin, David Ray (2000), page 101.

12   The division of human mind into scientific mind and mystic mind is an innovation of System Philosophy. But philosophers of pantheism tried to deal with metaphysical process using scientific reason.

13   This treatment of epistemology is my original idea. It is built upon the division of SPWV into rational part and empirical part.

14   Robert John Russell (Editor) (2004), pages 46 and 77.

15   Regarding the holographic paradigm of David Bohm for linking spirituality with quantum physics, see the two articles respectively of Kamladevi R. Kunkolienker and Richard D'Souza appeared in *Omega – Indian Journal of Science and Religion* (Institute of Science and Religion, Aluva), volume 7, number 2 (December 2008), pages 104 – 137.

#16   Wikipedia says (as accessed on 20 August 2017):

Systems philosophy was founded by Ervin Laszlo in 1972 with his book *Introduction to Systems Philosophy: Toward a New Paradigm of Contemporary Thought*. "Systems philosophy", in Ervin Laszlo's sense of the term, means using the systems perspective to model the nature of reality, and to use this to solve important human problems. Laszlo developed the idea behind systems philosophy independently of von Bertalanffy's work on *General System Theory* (published in 1968), but they met before *Introduction to Systems Philosophy* was published and the decision to call the new discipline "systems philosophy" was their joint one.

17   Brooke Noel Moore and Kenneth Bruder, (2005), page 161.

18   Kant, Immanuel (2003), *Critique of Pure Reason*, translated by J. M.D. Meiklejohn (Dover Publications, New York, 2003), Transcendental Dialectic, Book II, Chap. III, Section IV, pages 331-336; See also Hick (1994), page 18.

19   Ewing (1994), pages 26 -30

20   The main references used for Eastern Philosophy are Grayling A C (Editor) (1998), Hiriyanna M (1932), King (2000), Pulingandla R (2005) and Radhakrishnan S (1931)

Chapter 5

System Philosophy of Reality

****

5.1 Dilemmas about Reality

5.2 Three Levels of World as a System

5.3 System Model of Ultimate Reality

Author's main original ideas are marked by [*].

The mark [#] gives the number of note at the end.

****

Within the hitherto parts of this book, we have discussed about the various doctrines of western philosophy and its application for the understanding of physical world, life and mind. So far our focus was on the area of epistemology, which is one of the two principal branches of philosophy. Now it is the right time to consider the other branch namely *ontology* or *theory of reality* that deals with the question: what is the reality or fundamental stuff of the universe? This presentation will be built upon the premise of the contents of second chapter fortified with the inputs from the subsequent ones.

We can realize at the outset that the eclectic term *reality* gets only very scanty mention in the books and dictionaries on philosophy as well as science and religion; such omission is due to the controversies and conflicts between the various conceptions about reality. Of course, we commonly use the word *reality* in the most abused and confused manner during various situations. Most of us talk about different kinds of realities, rather than *the* reality singularly.[# 1].

5.1 Dilemmas about Reality

Let us begin our deliberation about reality by reiterating the role of philosophy in the landscape of human knowledge. Science, religion and art as well as certain combinations of these subjects are the macro areas of knowledge, pertaining to the study of the natural objects of universe in different ways. Hence, these subjects are qualified as *first order knowledge*. In contrast, philosophy is called the *second order knowledge* because it aims to unify the different first order disciplines.

Here it is important to clarify the meaning of *reality* by contrasting it with term *phenomenon*. We can define phenomenon as any object that depends on another object through cause-effect relationship. On the other hand, **reality is the original cause of all phenomena taken as a whole**. Accordingly, reality is self-caused, infinite and permanent. The terms like *ultimate reality* and *ultimate truth* are commonly used as synonyms of reality. [# 2].

The main issue with the notion of reality is whether it agrees with the religious belief in God. For a believer, there is no reality other than God. To aggravate the confusion, there are different definitions of God in various religions. Moreover, the concept of God causes many philosophical dilemmas, as already explained in the previous book. In this context, the spread of atheism – the argument that there is no God – is a serious matter to be tackled. Hence a sound theory of reality must advance a secular view about the original cause of universe and, at the same time, somehow link with the disparate definitions of God.

Based on the above points, we want to develop *ontology* or *theory of reality* by asking the four questions:

- ❖ Can we conceive reality as a being (substance) or process having existence?
- ❖ What is the method of knowing reality?
- ❖ Specifically, can we know reality from the features of phenomenal objects?

❖ How do we link the notion of reality with the religious definition of God?

We have already explained that ontology (theory of reality) begins by accepting the fact that the stuff of our universe has dual parts – **matter and consciousness**. The first aspect to be considered is the reasons for the diverse notions of reality proposed in the history of western philosophy. For that purpose, the discussion in chapter 3 about six world views and seven theories of reality must be revisited. Subsequently, in chapter 4, we have described the various philosophical doctrines proposed so far. The following two tables can summarize our hitherto discussion.

Table 1: Six Worldviews [# 3][*].

	Content view	**Process view**
Rational view	*(W1)* Organic Worldview for knowledge of value (OWV)	*(W4)* Spiritual Process Worldview- Rational (SPWV-R)
	(W2) Mechanistic Worldview-Rational (MWV-R)	
Empirical view	*(W3)* Mechanistic Worldview-Empirical (MWV-E)	*(W5)* Spiritual Process Worldview- Empirical (SPWV-E)
		(W6) Physical Process Worldview (PPWV)

Table 2: Seven Theories of Reality [# 4][*].

	Content view	**Process view**
Rational view	*(R1)* Idealism or Theism	*(R4)* Rational Spiritual Process
	(R2) Deism	
Empirical view	*(R3a)* Materialism	*(R5)* Empirical Spiritual Process
	(R3b) Intelligent Design Argument(IDA)	*(R6)* Materialist Process

A special mention may be made here about **Intelligent Design Argument,** which has been described in the previous book *Life and Mind*. We have noted the fallacy of intelligent design argument; it need not be repeated here. However, a section of scientists who have theist orientation, including Stephen Hawking and Paul Davies, uphold the belief in Intelligent Designer in order to explain the nonphysical aspect – purpose, creativity and freedom – in the evolutionary history of physical world. But this idea carries the baggage of scientific realism; we have earlier presented the counter arguments for refuting realism of any kind. At this stage it is stated that the intelligent design argument is not acceptable because it is a curious mixture of theology and scientific realism.

The idea of Designer is an Inference to the Best Explanation (IBE), which is associated with scientific realism. We can note that this view suffers from the problem of induction and also skepticism. It is a great puzzle to explain the nonphysical aspect coherently. More over the notion of Intelligent Designer is a vague concept because it does not include any worshipful attribute of God. We can treat it as the empirical version of deism - the terms like *intelligent* and *designer* are mere descriptive usages without any ontological assumptions. Hence there are sound reasons for rejecting the Intelligent Designer Argument.

We have introduced the notion of **monistic philosophy** to describe the diverse streams of thought in the history of western world. This nomenclature is an innovative idea of this book for distinguishing the erstwhile philosophy against the newly proposed System Philosophy.

In this chapter we want to remove the ontological controversies through the insights of **System Philosophy** pertaining to physical and biological phenomena. This new principle of synthesis would serve as the summit of our philosophical quest. [# 5].

5.2 Three Levels of Natural Systems

We must follow the path of secular thought for combining rational and empirical views about reality so as to explain the aspects of life, evolution and mind. Hence, our plan for synthesizing the competing theories of reality is based on the following conditions:

- Reality must have the components called matter and mind (consciousness), since it creates the phenomenal things with the dual aspects of matter and mind. Further, we must be able to explain the hierarchy of things in the world.
- Religious notions of reality like theist God as well as the scientific concept of matter are phenomenal ideas pertaining to religion and science respectively. We need the definition of ultimate reality as the synthesis of religion and science.
- **Reality must be conceived in third-person perspective through secular philosophy under content view; then only reality can be said to exist.**

In the following sections, we will make use of the innovative idea of *system*, which is the central part of System Philosophy, in order to develop a comprehensive vision of reality.

The term **system** has been defined philosophically in the previous book *Life and Mind*, as a productive structure formed by opposite entities. The X-axis and Y-axis pertaining to analytical geometry are used to represent the opposite forces of productive structure similar to a factory. In economics, the structure called factory uses the dual inputs *capital and labor* for producing certain outputs; there are different levels of outputs or products depending on the technology of combining capital and labour. The *analogy of factory* is highly suitable for defining a system. We can note that the opposite entities called X and Y separately cannot be conceived independently and absolutely; they are complementary aspects which combine to form a whole like factory. The production function model of system can be applied to both *phenomenon* and *reality* depending on the context.

It is necessary to explain how the system model shows existence, without attracting the problem of realism. It has been shown earlier in this book that our knowledge is produced by the sequence of theory (Ty), hypothesis (H), deduction (D), testing (T) and inductive inference (I) – it is denoted by TyHDTI scheme. Here Ty, H and D together form the group of deductive propositions (DP), while T and I together form the group of inductive propositions (IP). In other words, knowledge is the combination of DP and IP. As a rational principle, the system model is included in theory-part and it is a predicate. We cannot attribute existence to the predicate, as argued by Kant. But the system model can be verified by scientific evidence from worldly things in accordance with the TyHDTI scheme of knowledge. So we get the inference that the system model has existence. In other words, **"system exists" is a true inference, without involving realism or metaphysics.**

At this stage we have to develop a comprehensive view about the three levels of nature - *inanimate world, biological world and mental world*. For this integrative treatment, it is necessary to resort to the System Philosophy of biological organisms and human mind, as presented earlier.

Through ordinary perception we think that the **inanimate world** is physical and it is made of matter and energy. But there are serious dilemmas about the early stage of physical world as well as about the existence of matter. We have discussed these cosmological issues in the earlier book, *Origin of Universe*. Most importantly, the scientific realism and pluralism, which are allied to materialism, have failed in suggesting a coherent vision of cosmology. In this predicament, System philosophy offers a plausible solution.

We have already proposed the innovative *system philosophy of science* whereby the physical world has phenomenal existence on account of matter-energy duality. Accordingly, dark matter and dark energy exist as X coordinate and Y coordinate respectively and hence, they jointly form a system. As per the system model, the productive relation between X and Y produces various levels of products such as elementary particles, basic forces, atoms, molecules, galaxies, stars and planets. The system of dark matter and dark energy can be treated as parental universe having ten or more dimensions.

The event called *big bang* is the beginning of our universe on account of the compactification of a part of previous universe, so that the extra dimensions are hidden in order to manifest the four dimensional matter-energy world. In this view, our universe (physical world) grew from a singularity point by continuous transformation of some parts of previous universe. This process can be compared to the growth of a seed to become a tree by drawing resources from the environment. It may be emphasized here that matter and energy have complementary existence in the system model of physical world. So we have to discard the notion that matter is an illusion. All physical things exist phenomenally as matter-energy systems. Note that matter and energy are not separate substances; they have complementary and interdependent existence as opposites.

The system model of physical world serves the purpose of justifying the scientific laws. Science basically deals with physical things made of matter and energy; subsequently it reduces living beings into material bodies. We can call this as the *process of physicalisation*, which is guided by the practical objectives of science. However, we can realize that this approach does not explain the nonphysical aspects of creativity and purpose observable even in the physical world. How can we account for the hierarchical order of complexity achieved by the evolution of world in the order of cosmological stages? Atoms combine variously to form different molecules so that new properties emerge wonderfully.

A typical example is the formation of water when hydrogen and oxygen combine in a certain proportion. The emergent properties of water are strikingly at a higher level than that of the constituents; the formation of water upon earth is the beginning of a long chain of phases that caused the origin of life in the oceans about 3.5 billion years ago. Evidently, there is a nonphysical aspect – purpose, creativity and freedom – in the evolutionary history of physical world. It is a great puzzle to explain the nonphysical aspect coherently. Moreover, we do not want to resort to the *intelligent design argument*, due to its inherent philosophical problems. Happily, our problem will be solved through system model incorporating consciousness for representing the nonphysical aspects.

System philosophy clarifies now that the system model of physical world is a practical and phenomenal principle for justifying scientific laws. In order to add the property of nonphysical aspect we must conceive of a higher system formed by matter and consciousness, which becomes the physical world by reduction into matter and energy. It is expedient to denote the said opposite forces as **matter1** and **consciousness1** respectively, which form a system – this is the basic level of natural world. Let us see how this idea is applied to a factory producing car. Through suitable mathematical techniques, the labour and capital can be transformed to matter and energy respectively. So, as per the analogy of factory, labor-capital system is first transformed to matter-energy system and then to matter1-consciousness1 system.

Next we will consider the **biological world** that is the second level of natural world. The system model of biological world shows that living organism exists as an X-Y system of macromolecule and information, as explained earlier. We may give the important points in this regard for better understanding.

The central idea in modern biology, which adopts the physical process view, is that *life* is a physical process guided by the DNA and genetic code present in the cells of organism. In this way, it is proposed that organism is like a computer – the molecular structure of cells and its higher organizations constitute the hardware while the genetic code serves as the software (algorithm). Through philosophical arguments we can show that the machine-algorithm model lacks justification. The existence of DNA and other macromolecules cannot be established on account of the deeper problems about the notion of matter. Mainly, the existence of separate entities called DNA and genetic code cannot be proved, since they are predicates as per the philosophy of Immanuel Kant. Moreover, genetic code displays nonphysical aspects of creativity, freedom and purpose. In this milieu, the existence of DNA and genetic code as physical objects suffer from grave inconsistencies. [# 6][*].

In order to tide over this critical problem, we must apply the *system model* and hold that life is a system formed by the opposite entities called macromolecule and information. The phenomenon of life has non-dual existence in the form of a system. It is due to the influence of the Aristotelian rules of thought that we conceive macromolecule (mainly DNA) and genetic code as separate objects. It has been already established that the information contained in genetic code is nonphysical. We propose to link DNA and information by treating them as X and Y axes respectively. So we rename macromolecule and information as **matter2** and **consciousness2** respectively.

Further it may be emphasized that it is a religious thought to treat life as a formless being that exists in physical body. At the same time the scientific theory that life emerged from physical DNA also is a fallacy. System philosophy of life does not resort to the notion of intelligent designer for explaining the nonphysical aspects of life – creativity, purpose and freedom – exhibited by cell and its higher organizations. Thus, finally, the puzzle of life is solved here.

Next step is to take up the philosophical treatment of biological evolution. Only content view can address the concerned issues since the process view gives only the description of the historical stages of evolution. *So we may switch over to the principle that evolution must be perceived as a phenomenon under content view of knowledge.* Now we proceed to find the method for synthesizing naturalism and vitalism in a secular manner. As a continuation of the above system model of life, we can show that the dialectical and productive relation between X and

Y will produce different levels of organisms with increasing complexity. Thus we get the diagram called *the system model of biological evolution*. It may be emphasized that this model is the correct alternative to Intelligent Designer Argument. [# 7][*].

Finally, we assert that Darwinism is a physical reduction of the system model adopting the process view. The nonphysical aspect of information or consciousness is reduced to a mechanical algorithm called genetic code. Then DNA and genetic code together constitute the form of machine-algorithm, which is the key principle in the physical theory of Darwinism. Explaining organic evolution in physical terms serves the pragmatic objectives of science though it fails philosophically. We emphasize that Darwinism will continue in use for the scientific study of evolution. But it raises serious conflict between science and religion.

Thirdly we want to propose the system model of **mental world**, especially the existence of human mind. The prevailing scientific theory about human mind is *Computer model functionalism* – it holds that nervous system functions like a computer. The neuron networks form the hardware. The mental states are like the software and it denotes the function of neuron networks. The machine-algorithm view of computer is used by neuroscientists. Through philosophical arguments, it is possible to show that the computer model of mind is defective. System philosophy appears as the plausible explanation about the existence of the nonphysical structure of mind.

As the first step towards the new philosophy of mind, we assert that diverse mental states – feelings, emotions, ideas, memory etc. – are produced when a higher form of consciousness works on the neuron signals of *brain and other parts of nervous system (BNS)*. It may be noted that BNS is a part of biological body. Let us denote the BNS by *matter3* and the higher form of consciousness as *consiousness3*. In this way we can note that **mind is a matter3-consciousness3 system.** This dialectical production model illustrates the existence of various mental states in the static way of content view.

From this perspective, it is now clear that human being exists as a system with three subsystems -- inanimate system, biological system and mental system. This theory can be extended to the nature as a whole, as shown below.

Table 3 : Three Levels of Nature [# 8][*]

Three Worlds	Theoretical Entities (Predicates)	Phenomenal Existence (System)
1. Inanimate World	Fundamental opposite aspects within atoms (physical and non physical)	(matter1, consciousness1) system
2. Biological World	DNA, RNA, protein and other macromolecules with genetic code	(matter2, consciousness2) system
3. Mental World	Neuron signals and non-physical aspects in BNS	(matter3, consciousness3) system

System philosophy of mind gives the *happy solution of the body-mind dualism* which is arguably the greatest riddle of philosophy. It is our custom to say that mind is a nonphysical structure that exists above the physical neuron networks. This kind of separation of mind from nervous system is a consequence of our ordinary thinking in accordance with Aristotle's rules of thought. Now we can replace the body-mind dualism with body-consciousness system. [# 9][*]

5.3 System Model of Ultimate Reality

For a successful theory of reality, the following three conditions are to be satisfied.

- We must define reality without assuming realism. We may remember that realism asserts the separate existence of matter and mind in accordance with Aristotle's rules of thought.

- The monistic notion of reality led to the six worldviews and it must be synthesized taking into account the opposite conceptions.
- Existence is the aspect of permanence and it has meaning only in the content view of reality. Process view is to be applied for describing the aspect of change. So reality has the dual properties permanence and change.

You may ask: since the word *reality* means "that which exists really", how can we avoid realism? Is it possible to remove realism that is a jump from the abstract definition of class to the notion of existence? There are occasions where we talk about the existence of things about which there is no empirical evidence. The statements about God, soul, angel and heaven would fall into this kind of propositions. When I say that such theoretical entities exist, it attracts the problem of realism.

Consider the sentence "a table exists in front of me", it means that I have correct phenomenal knowledge as per the TyHDTI scheme about an object called table. It is an inference proved by sensory evidence. Now we can modify the position of realism by asserting that **the existence of a thing is an inference** known through the said scheme. That is, we cannot go beyond our scheme of knowledge to say that a thing exists as external to our mind.

It is now clear that propositions about reality also must be formulated through the TyHDTI scheme of philosophical knowledge. Here, the definition of reality as the original cause of universe would belong to the theory part. Accordingly, reality is a theoretical entity which is given meaning through the evidences obtained by observing the objects of universe in an integrative manner.

In the above Table 3, we have introduced the subscripts 1, 2 and 3 to matter and consciousness for distinguishing the three levels of phenomena. And it also implies that the fundamental constituents of universe are matter and consciousness, since they manifest in the hierarchical levels observed. For defining the structure of ultimate reality we may replace the word matter with Body and treat consciousness in the real sense. Thus we get the definition: ***Ultimate Reality or paramporul*** *is the system of opposite forces called **Body** and **Consciousness**.* Representing Body by X-axis and Consciousness by Y-axis, we can illustrate the production of inanimate, biological and mental worlds – it is called the ***System Model of Ultimate Reality.*** Alternatively we may call the diagram as *System Model of Nature* or *Picture of Ultimate Reality* also. [*]

The above method may appear as an adaptation of logical positivism to the realm of philosophical theory. Adhering to the verifiability criterion of meaning, the system model of ultimate reality is correct knowledge because it explains the fundamental aspects of matter, mind and evolution. The X-Y model is a *theory* presented in content view; it is verified using the evidences about natural systems. Subsequently, we make the inference that reality exists simply in the form of a factory. That is, the truth of the model is ascertained by empirical evidences about the hierarchy of things in the cosmos.

We have now reached the final stage of our philosophical search; it is the vision about the ultimate reality, which is axiomatically defined as the first cause of all phenomena as a whole. The oriental term ***paramporul*** can also be used for denoting the ultimate reality. This ontological principle necessarily expresses the synthesis of the three levels of phenomena in nature – inanimate world, biological world and mental world. Though the existence of ultimate reality is my *inference* supported by evidences as per the TyHDTI scheme, I can practically believe that ultimate reality has real existence by the strength of the ontological arguments given below. [# 10][*]

1. I think, therefore I exist as a system of matter and consciousness.
2. All phenomenal things, which served as the cause of my being, exist as systems.
3. Therefore nature exists as a phenomenal system.
4. I am practically justified in believing that Ultimate Reality exists as a system to produce all phenomena included in nature.

In the foregoing paragraphs, we have presented the **essential structure** of ultimate reality. This deductive method can dispel the doubts of all confused minds who ask: Is ultimate reality knowable? Can we conceive the existence of ultimate reality through content view?

The X-Y coordinate system has four quadrants (or parts) denoted by I, II, III and IV. When we consider the hierarchy of phenomenal things in a factual manner, without invoking the value aspect, all things can be shown in quadrant I using the graphs called isoquants. It may be clarified that isoquants are special kind of graphs employed in economics for representing the separate products of a factory. The analogy of a factory for producing car is the best key to understand the function of Ultimate Reality. In the factory, elementary components are joined to make more complex components in successive stages to achieve the final production of car. In the similar manner, we can describe the production of various inanimate and living things of natural world.

Now suppose that we judge a phenomenal thing as good or bad under the norms of ethics. Then good thing will be placed in quadrant I, while bad thing is in quadrant III, of the System Model of Ultimate Reality. Generally, we postulate that phenomenal universe occurs in quadrants I and III; these divisions stand for good and bad respectively. The logic for distinguishing between good and bad phenomena is that Ultimate Reality possesses the property of purpose or value which has positive and negative directions. The idea of dividing value (purpose) into good and bad aspects may sound new to the readers who habitually identify value with the good aspect only. The system model shows symmetry in all respects.

The principle that Ultimate Reality exists as a system with good and bad parts explains the occurrence of good and bad aspects in phenomenal world. In other words, these opposite qualities have ontological existence as per the system model. We can rightly observe that when goodness increases, bad features also will increase.

In this context, it is important to realize that the system model can be interpreted according to process view also. Matter and consciousness are in dialectical and productive relation causing the historical occurrence of phenomenal things. Another advantage of

System Model of Ultimate Reality is that it removes the confusion caused by Intelligent Design Argument. Many of the Christian theologians and scientists have subscribed to this argument thinking that it expresses the relation between God and natural world through empirical approach. This way of linking theology and science suffer from serious philosophical inconsistencies. We may add some pertinent points in the following paragraphs.

Reality and God

The definition of reality has been an intractable problem in hitherto philosophy mainly due to two reasons:

> ➤ The metaphysical realism, which is a consequence of Aristotle's rules of thought, paved the way to the religious idea of God as ingrained in the psych of believers.
> ➤ The search for an alternate definition of reality has been hampered by religious belief. It resulted in the traditional conflict between idealism and materialism.

Under System Philosophy, the Ultimate reality is the original cause of the opposites contained in all kinds of phenomena – scientific, religious and artistic – appearing through content view. Also, as we have stated earlier, the true nature of philosophy is *superscientific*, in the sense of secular and logical inferences, meant for synthesizing all types of phenomenal knowledge.

Ultimate Reality is the original cause of the production of hierarchical order in universe, culminating in the emergence of human beings with matured brain and mental faculties. In human mind, its unconscious part is the source of purposes and motivations, which are responsible for the formation of our social systems. We may analyse the resemblances and differences of the multitude of social systems developed by human species. In this way we can classify the entire spectrum of manmade social systems into *seven macro systems* named as natural life system (NLS), economic life system (ELS), political life system (PLS), family life system (FLS), ethical life system (ETLS), religious life system (RLS) and artistic life system (ALS). The system model of human mind shows that the seven life systems mentioned above are produced by separate faculties of human mind, according to the respective aims of practical life. Next chapter will elaborate this theory.

The justification of religious propositions – mainly the statements about God, soul, angel, evil, hell and heaven – can be established using the above definition of religious life system (RLS). This topic has been explained in the previous book *Life and Mind*.. The most challenging issue in religious belief is that there is no proper explanation about the origin of evils in the universe. This theological question is called as **problem of evil**, which troubled thinkers throughout the history of philosophy. We may now state that the problem of evil is essentially connected with religious definition of God and it is expressed in the following question. How can a perfect and good God create a world in which there are various kinds of evils? How can God allow evils to happen? Surveying the ideas of great philosophers, S.E. Frost writes:

"God is thought as the source of all good and the devil is the evil principle. But, to the question "Did God create the Devil?" there is no answer. A dualism of good and evil works well until the attempt is made to account for the creation of the universe; but that presents difficulties which have not been solved". [# 11].

The issue of evil actually belongs to the topics of religion and ethics. In the present context, it may be reiterated that the *System Model of Ultimate Reality is radically different from the religious notions of God, Devil and Soul*. As per the teaching of religion, God is regarded as the reality, while other entities like evil, soul, mind, and matter have subsidiary existence only. This idea of religious reality serves as the theory of religion, which is the product of our mystic faculty of mind. *For the time being, it is enough to say that the System Philosophy does not negate the religious ideas of God and Devil*. We will discuss the truth of religious propositions in the final chapter. Thus we would address the question as how to link the notion of reality with the religious definition of God. [# 12][*].

Certain interesting corollary of the foregoing theory may be mentioned with regard to the traditional doctrines of spiritual process. Neo-Platonism, the oldest form of spiritual process doctrine, calls God as the One. Philosophies of Spinoza, Leibniz and Hegel held that the reality is a mysterious union of matter and consciousness from which the phenomenal opposites emanate. We can reinterpret this idea using our system model in which *the point of origin* of X-axis and Y-axis represents the spiritual process notion of reality. Obviously this queer idea needs modification for showing existence under content view. [# 13][*].

Additionally, the relation between Brahman and Maya, pertaining to Vedanta philosophy of Hinduism, is a serious matter of confusion. Concerned theologians say that the Brahman is the reality and it exists eternally together with Maya power. By converting this statement to the language of system model, it implies that Brahman and Maya together are situated in the original point of the X-Y coordinate system. We can now remove the philosophical inconsistencies by suggesting that Maya is the X-axis while Brahman is the Y-axis; then reality is like a factory process with opposite components of infinite measure. From the forgoing, we may reiterate that the philosophical issues of all spiritual process doctrines of west and east can be removed by using the system model of matter and consciousness.

NOTES of Chapter 5

1   Most of the important academic books on philosophy have avoided specific mention of the term *reality* in the sense of original cause. A helpful explanation of the concept of reality is available in George Thomas White Patrick (1978).

2   This chapter is in the form of conclusion of the philosophical discussion covered so far. ***Hence the reference books already mentioned in the earlier chapters hold good for the present chapter also.***

#3   It may reiterated that the definition of six worldviews and Table 1 constitute the basic framework for developing System Philosophy.

#4   The positions of individual theories of reality in Table 2 would indicate the method of unifying the seven theories of reality. The point highlighted here is that theories of reality are derived from worldviews.

#5   Project of System Philosophy is to show that the three levels of natural world – inanimate things, living beings and mental phenomena – are phenomenal systems of matter and consciousness. These systems are originally caused by the Ultimate Reality that also is a system.

#6   The nonphysical aspects of genetic code and the definition of system have been initially presented in the previous book *Life and Mind*. It is practically easier to explain the concept of system when we discuss the purpose and creativity in life and evolution. This idea is the spring board of System Philosophy.

#7   It is my original idea to conceive evolution from content view. The intelligent design argument has been explained critically in the previous book *Life and Mind*.

#8   By treating the three levels of nature as systems, we can incorporate the nonphysical aspect of purpose. This is my original idea.

#9   The concerned points about body-mind dualism have been explained in the previous book *Life and Mind*.

#10   The arguments given here form a radical modification to the axiom of Descartes - *cogito ergo sum* (I think, therefore I exist) – about the separate existence of self or mind in contrast to body.

\# 11   Frost. S. E (1989), *Basic Teachings of the Great Philosophers,* page 91.

\# 12   System Philosophy about God and Evil, which is my original idea, has been introduced in the previous book *Life and Mind.*; a separate volume on Philosophy of Religion is required to elaborate this topic.

\# 13   Comparison of pantheism (spiritual process) with System Model is my original idea.

Chapter 6

Seven Life Systems and Knowledge

****

6.1 Social World and its Components

6.2 Existence of Seven Life Systems

6.3 System Model of Knowledge

Author's main original ideas are marked by [*].

The mark [#] gives the number of note at the end.

****

Now we move on to study the human actions that constitute social systems. *We want to shift our philosophical analysis from natural world to social world.* This new focus is vitally important in the context of enquiring about the good and bad aspects of our voluntary actions as well as behavior. At the outset, the hierarchy of various groups and societies will be explained. This will lead to the notion of institution. Then we can introduce the innovative idea of *seven life systems;* these are non-overlapping institutions at the global level.

The seven life systems exist depending on the structure of human mind. So the *system philosophy of mind and ultimate reality* provides the *final justification* for the laws and inferences of knowledge under various disciplines. So it will pave the way for the epistemological synthesis of the entire spectrum of human knowledge.

6.1 Social World and its Components

It can be observed that there are certain patterns in the social behaviors of people, though it can vary according to different periods and situations. Accordingly, different levels of human groups are formed, which together constitute the **social world**. It includes the entire range of actions performed by persons when they are in groups. In order to develop our understanding about social world, we must propose a suitable taxonomy (classification) of its components.

The fundamental unit of social world is a **group**, which consists of two or more inter-connected persons. A group has the following vital characteristics.

- C1) Members share some common goals pertaining to the social relation. In other words, the group has certain *norms* and *culture* (these concepts will be defined below).
- C2) Members interact with one another in accordance with the norms and culture of the group.
- C3) Members share the rights and obligations of membership. Hence every member has a particular *role* or *social status*.
- C4) Members have a sense of identity and belonging within the group.

We can introduce the concepts called norms, culture and role with suitable definitions. Every group has a particular pattern of social behaviors; this is because the members accept certain norms. Hence, the word ***norm*** stands for the particular social rule or value which account for the uniformity of behavior of members of a group. The dress code of nurses in a hospital and that of girls in a school, are examples of such norms. It may be added that the members of a group get adapted to the concerned norms. The totality of rules and values forms the ***culture*** of the group. Culture is an acquired quality, which is learned through socialization and habits. Culture is communicated from one generation of members to the next generation. Moreover, culture can undergo change depending on the needs of group. However, it is a moot point whether the influence of culture on the social behaviors of a group or society can be studied satisfactorily through scientific methods.

Thirdly, we can define **role (social status)** as the particular function of a member within the group. Hence role refers to the actual behaviors expected of a person because of the position he or she occupies in a social structure. Note that father, son, mother, husband, wife and daughter are the important roles within the group called family. A group exists as a pattern of social relationships between the roles, rather than individuals. The social structure of a group is constituted by the concerned roles of members.

We have mentioned that the members of a group have a common set of goals which give rise to the norms and culture of the group. It is possible that there are many groups with the same goals; but such groups may be separated on the basis of geographical area or function. In a school there are many groups of students who are belonging to different standards as well as disciplines. Similarly we can talk of different groups of farmers in a country. Considering the family resemblance of goals, there are different classes of groups that form a *hierarchical order* -- namely community, association, social organization and society. It is necessary to present a definitive idea about the higher combination of groups as below.

The term *community* stands for a natural group of individuals living in a given area. The tribal community in a particular area of forest is the best example. The families serve as the primary groups in the tribal community.

Association is commonly interpreted to be a large group or a union of many groups with specific purpose and set of rules. A political party or trade union can be cited as example. In the field of Christian religion in Kerala, we can treat various churches (including lay members and priests) as groups; then diocese is to be considered as an association.

Social organization is the structure of two or more associations which are interrelated. It has a formal structure in order to achieve some specific goal or goals. There are specialized tasks for the members, to be coordinated by the leaders of the organization. Hence the structure of social organization is hierarchical. Examples: in the realm of economic activities, the main social organizations are banks, companies, agricultural producers (including wholesale dealers in the markets), government and consumers. Regarding the Christian religion of Kerala, the Syro-Malabar sect is a social organization, which is the union of certain number of dioceses.

Next we may explain the concepts of *society* and *institution*. These are rather abstract notions, and there is no unanimity about their definitions. Here we follow the scientific approach of social sciences, especially sociology, for defining these concepts.

The most important views about **society** are given below.

- ❖ Society is the pattern of relations between a set of social organizations which share a common purpose.
- ❖ Society is the largest possible group of human beings who are interconnected by a particular pattern of social relationships. Thus, society is not a sub-group of any other group.

These disparate definitions are exemplified by a wide spectrum of societies depending on the specifications of geographic area and functional goal. For illustrative purpose, the following are the societies considered for empirical study:

1. Tribal society (at the global level).
2. Consumer society …
3. Urban society …
4. Economy of Kerala / India / China.
5. Indian Society, considering the cultural aspects
6. The political party called Indian National Congress

From the above list, we may affirm that geographic area is not a significant criterion for defining a society. The economies of Kerala and India are distinct societies, as per the method of sociology, though the former is geographically included in the latter. The pattern of social relations in a society is viewed as a whole in functional way. It may be clarified that the separate societies are distinguished on the basis of the goals, purposes and culture of the concerned human beings who engage in social relationships. This fact leads us to the articulation of the concept of *institution*.

As explained earlier, society is formed by individuals through the hierarchical stages starting from groups. Hence group or society is a super structure existing above the set of individuals. The principle is : *whole is more than the sum of parts*. So it is reasonable to ask: what is the reason for the formation of group in the first place? Why do individuals come together in order to form a group? Sociologists adopt the notion of institution to answer these vital questions. They propose that individuals engage in social relationships because they have certain goals or values derived from various institutions. We may define **institution** as the set of universal rules which serve as the source for the establishment of a particular group or society. In other words, institution is an organized form of customs, dogmas, rituals and procedures that preexist before individuals enter into social relationships.

The meaning of the term *institution* in the sociological parlance is different from our ordinary usage, where we say that a school or hospital is an institution. Here it is expedient to distinguish between society – including the sublevels such as group, community, association and social organization – and institution. Every society is a concrete set of persons engaged in a form of social relationships. On the other hand, institution is an abstract structure of values, goals, traditions and rules, which is concerned with one particular aspect of social life. Viewed in this way, we can say that the main institutions are family, marriage, educational institution, religious church and hospital, when these terms are taken in the general or universal sense. In contrast, family of Gandhi is a group that is the sublevel of the society called families of India.

The phrases **'social system'** and **'social institution'** are very popular in the context of describing societies and social phenomena. As an ordinary usage, *social system* refers to a society or its sublevel. But in a technical way, we can give a more precise definition as following: When societies of a particular class are aggregated up to the global level, the totality may be called a *social system*. Different social systems are formed on the basis of institutions such as economy, family, religion, art, polity (state), ecosystem and ethnical system, when considered for the whole world. Hence *institution* denotes the set of goals and values expressed in a particular subset of a social system.

Institutions are of paramount importance in the practical life of human beings. William C Levin, in his book *Sociological Ideas*, holds that the universal goals pertaining to various institutions are procreation, sexual access, care of the young, socialization, distribution of power, social control, production-distribution-consumption of goods, education and religious worship.. Accordingly he mentions that some major American institutions are family, Government (or Polity), Economy, Education and Religion. [# 2].

It is now obvious that *institutions* have fundamental role in our knowledge about various groups and societies. However, the reference books used for this chapter do not throw light on the question: **how did various institutions come to exist in the first place?** The scientific approach of sociology and other social sciences is helpless in this matter. We have to undertake the philosophical analysis for understanding about the origin of institutions. The following sections are designed with this objective.

6.2 Existence of Seven Life Systems [# 3][*]

It may be asserted that the key concepts of institutions were originated in the earliest stage of human civilization. Since an institution is a pattern of rules and customs, it is an abstract entity. In contrast, group or society is a concrete structure including human beings and their possessions, pertaining to a particular period of time. It can be said that every person is born into the mesh of social institutions. Such institutions have the dual aspects of permanence and change. The basic structure and identity of an institution is permanent, but its constituents can undergo change over time and situations. That is, institution has a dynamic existence; it evolves through human generations.

Why did institution originate and evolve along with the advancement of civilization? Sociology and other social sciences, which adopt the physical point of view, would fail to answer this question. We may mention the main reasons as below.

- Social science proceeds in the path of materialism and empiricism for explaining social phenomena. It describes institutions as social facts, but it eschews the essential aspects of mind or consciousness. We cannot account for our values or goals, if we fail to recognize the nonphysical faculties of human mind.
- Our consciousness has the nonphysical properties like purpose, creativity, and freedom. These mental aspects are outside the scope of science and social science, which deal only with physical properties of phenomena. In this situation, social science cannot explain the existence of institutions which are the aggregates of purposes or goals.
- The empirical approach of social science leads to skepticism about social reality. It means that the existence of institutions cannot be known. Through epistemological considerations, we can point out that the so called laws about social phenomena do not have justification. Moreover, theory of social science suffers from pluralism about the factual description of institutions, without any possibility of unification.

In the wake of the above critical points, we may present an alternative line of thought - the *System Philosophy of Seven Life Systems* - for classifying the multitude of institutions. Our attempt is to show the separate existence of institutions so as to unify them through the system model. For that purpose, the defining aspects of values (goals) have to be demarcated.

We can note that values represent our goals of life. Since the goals have good and bad directions, it must be admitted that the words denoting values can be arranged in pairs of opposites. Love-hatred, kindness-cruelty and courage-timidity are a few examples.

Every value, as it occurs in our mind, has dual aspects of self-interest (SEI) and society interest (SOI). For example, when I love my neighbor, it serves my selfish motives and at the same time satisfies the interest of society. In other words, the act of loving my neighbor is beneficial to myself as well as to the society. Similarly we can analyze hatred as having detrimental effect to me and society. That is, hatred contains the negative aspects of SEI and SOI. This shows that each of the polar opposite of a particular goal is a combination of SEI and SOI.

Now we reach at an **innovative method** for classifying the goals or values of our life. Firstly, the goals are separated into good and bad classes. Secondly, each goal is divided into the dual aspects of self-interest and society-interest. We can combine these two steps by taking self-interest as X axis and society-interest as Y axis, as per the coordinate system of analytic geometry. The positive sides of X axis and Y axis represent the good class while the negative side shows the bad class. *Accordingly, every goal is a system of self-interest and society-interest so that the good aspect of goal is the positive side and bad aspect of goal is the negative side. This is the system model of a goal.* It is our linguistic practice to use separate words for denoting good and bad aspects of a particular goal. In other words, the system model enables us to unify the conventionally opposite goals like love and hatred.

The next step is to apply the system model to an institution that is the aggregate of a set of goals having family resemblance. ***The institution exists as the X-Y system of self-interest and society-interest; this system has both good and bad aspects.*** Now it can be proposed that social institution originated from the particular faculty of human mind in the early stage of civilization and subsequently it evolved. To prove this point, let us recall the *system model of human mind* according to which the mind exists as a system of body and consciousness. Translating this model of mind into the model of self-interest and society-interest, we can hold that mind is divided into various faculties causing the social institutions. Strictly speaking, the faculties of unconscious mind are primarily responsible for the origin and existence of social institutions.

Through meticulous analysis and practical evidence, the multiplicity of social institutions can be unified by a process of aggregation up to the global level. So we reach at the enlightening thought that, for human race as a whole, there are seven non-overlapping social systems, which are appropriately called as **seven life systems**. They are like the seven continents upon earth. Or each life system can be compared to a kingdom - plant kingdom or animal kingdom - in biological world. The local level institutions are sublevels of these life systems.[*].

The names of the *seven life systems* are proposed as below:

1. Natural Life System (NLS)
2. Economic Life System (ELS)
3. Political Life System (PLS)
4. Family Life System (FLS)
5. Ethical Life System (ETLS)
6. Artistic Life System (ALS)
7. Religious Life System (RLS)

The *life systems* are global-level social systems, by virtue of four main reasons to be mentioned here in brief. Firstly, these macro institutions have similarity with the properties of life -- they originated at the dawn of civilization; they grow by drawing energy from the cultural environment; they have various adaptive mechanisms; they are mutually interacting and they have distinctive structures with hierarchy of internal levels of organization. Secondly, the term 'life' also signifies the fact that the span of human life is divided into the participation in these macro systems. For example, when a person purchases a shirt from a shop he is participating in ELS; when he performs religious worship, he is in RLS; and so on.

The term 'system' is valid on account of the X-Y model introduced earlier. We can say normally that each *life system* exists as a combination of body and consciousness, because it is a higher organization of natural things on the basis of human social behavior. But from the ethical view of human actions, life system is a system of SEI and SOI. It is like a factory for producing various groups, associations, organizations and societies, which generally have a particular type of goal.

In the context of a *life system*, the term 'goal' is usually used as a collective noun. As a matter of fact, there is a set of goals for the life system; these goals are interrelated so as to make a whole. We can observe a hierarchy in the goals when we compare between the various levels of groups. Moreover, the levels of goals are unified on the basis of the dual principles of SEI and SOI.

All these points indicate that the notion of *life system* is required for understanding the origin of a class of groups or societies, which manifest a specific set of common goals.

More importantly, there is a strong philosophical reason behind the idea of life system. Referring to the philosophy of Immanuel Kant (1724-1804), we can hold that the material and mental components of a social system are mere predicates if we consider them as separate theoretical entities. In that case, the separate components do not have existence by virtue of Kantian maxim that *existence is not a predicate*. It leads to the predicament that institutions (the aggregates of values or goals) cannot be said to exist; institutions are our mental constructions only. In order to tide over this skeptic issue, *we have defined life system as the dialectical relation between the opposites called SEI and SOI; this structure has phenomenal existence from our philosophic perspective. By virtue of this principle, we get justification for the laws and inferences about the concerned social phenomena.*

Having introduced the seven life systems, we can now complete the picture of the organic structure of our social world. There is a striking similarity and correspondence with the hierarchy in the biological world as shown below.

Levels of biological world - Organisms, Species, Family, class, phylum, and kingdom

Levels of social world - Groups, associations, social organizations, societies, institutions and seven life systems.

The most significant application of the above theory of seven life systems is to analyze the origin of social institutions. At the fundamental level, the numerous institutions are finally classified into seven life systems, which are further unified by the system model of

SEI and SOI. The other extreme is the level of individual members of a group or society where the members are like the organs of an organism.

A simple description about Natural Life System (NLS) may be given here for clearing the possibility of confusion. As a global level entity, NLS consists of the values or goals related to all aspects of natural world. Obviously, it includes inanimate things, living beings and mental phenomena. Specifically we can consider the values adopted in biological life pertaining to education, skill development, games, entertainment, food habits, shelter and clothing. The cultural practices pertaining to the birth of a child and funeral of dead bodies can also be included in NLS, though such practices have overlapping with FLS and RLS mainly. For pragmatic purposes, we can divide the social groups formed on the basis of NLS into three levels – inanimate world, living world and mental world. ***By physical reduction***, we get scientific knowledge about these worlds; this is the broad area of natural science.

At the present level of advancement of human society, it is very difficult to separate NLS from other six manmade life systems. However, when a person is stripped of the paraphernalia of the latter set, he or she still remains in NLS conducting life within the social network including related persons and natural environment. We can hold that NLS is the primary level of human life. It can be said naively that by NLS, we mean the values and goals pertaining to the way of human life as found in primitive communities as well as nomadic people. NLS is the basic form of civilized life, since the other six manmade life systems are secondary developments.

Sociology is the scientific study of social behavior of individual human beings. Human social behavior happens primarily in NLS; but in most of the cases, such behaviors occur at the intersection of NLS with one or more of other life systems. A few examples of social behavior analyzed by sociology are given below.

- Divorce, which in a phenomenon occurring in NLS and FLS.
- Unemployment, which is a phenomenon occurring in NLS, ELS and PLS.

- Dowry, which is a phenomenon occurring in NLS, ELS and RLS
- Sports culture, which is a phenomenon occurring in NLS, ELS and PLS
- Industrial strike, which is a phenomenon occurring in NLS, ELS, PLS and ETLS

The psychological dimension of social behavior, together with the relevant roles of people as members of various life systems, is brought to scientific analysis under sociology. The definitions of economics, political science, family science and ethics run in similar lines. These disciplines consist of scientific study of the structure of ELS, PLS, FLS and ETLS respectively. The salient features of these subjects – normal social sciences - are beyond the scope of this chapter. However, their contrast with sociology has been noted above.

Our particular mental faculty called *religious mind* motivates individuals to form religious groups. Accordingly, they perform a variety of actions for worshipping supernatural beings like God, different kinds of gods and venerable souls called saints. God is regarded as the supreme level of such metaphysical beings as per the belief that God created this universe. Through the history of human civilization, multiple forms of groups and societies were established with the aim of religious worship. The relatively more organized societies of this kind are classified as *religions* (like Christianity, Hinduism and Islam), where as smaller organizations are called cults. Religious Life System (RLS) is the global level institution representing the totality of religious societies having various levels of organization. RLS has existence due to the duality of SEI and SOI inherent in religious worship.

In similar lines, we can hold that Artistic Life system (ALS) exists with the fundamental goals of experiencing the beauty in various forms of art. For example, the social relationships between a set of publishers, writers and readers can be treated as a group in the field of literature. Evidently there are numerous groups and societies with organizational structures of great variety for the values of aesthetic enjoyment. The key point of system philosophy in this context is that there are opposite goals of SEI and SOI when a person engages in the field of art, either as artist or audience. The benefits accruing to the individuals (the self-interest aspect) can be listed easily. It may be clarified that the society interest in this regard is reflected in the feeling of harmony and togetherness among people.

The notion of *life system* enables us to understand societies and social systems using philosophical and ethical principles. It is instructive to add that in majority of cases, a society draws the values from two or more life systems. The best example is a society involved in terrorism that has at least political (legal), economic and religious dimensions. When all terrorist societies are aggregated to the global level, this totality can be treated as a social system. For justifying the empirical laws about terrorism as a social system, we must take into account the existence of the concerned life systems. So we can realize that *most of the social systems are intersections of two or more of the seven life systems*. **As per system philosophy, the social reality consists of the seven life systems,** which have existence for producing various kinds of institutions, societies and social systems.

Chapter 6 / Diagram 1
Seven life systems, many institutions and social systems

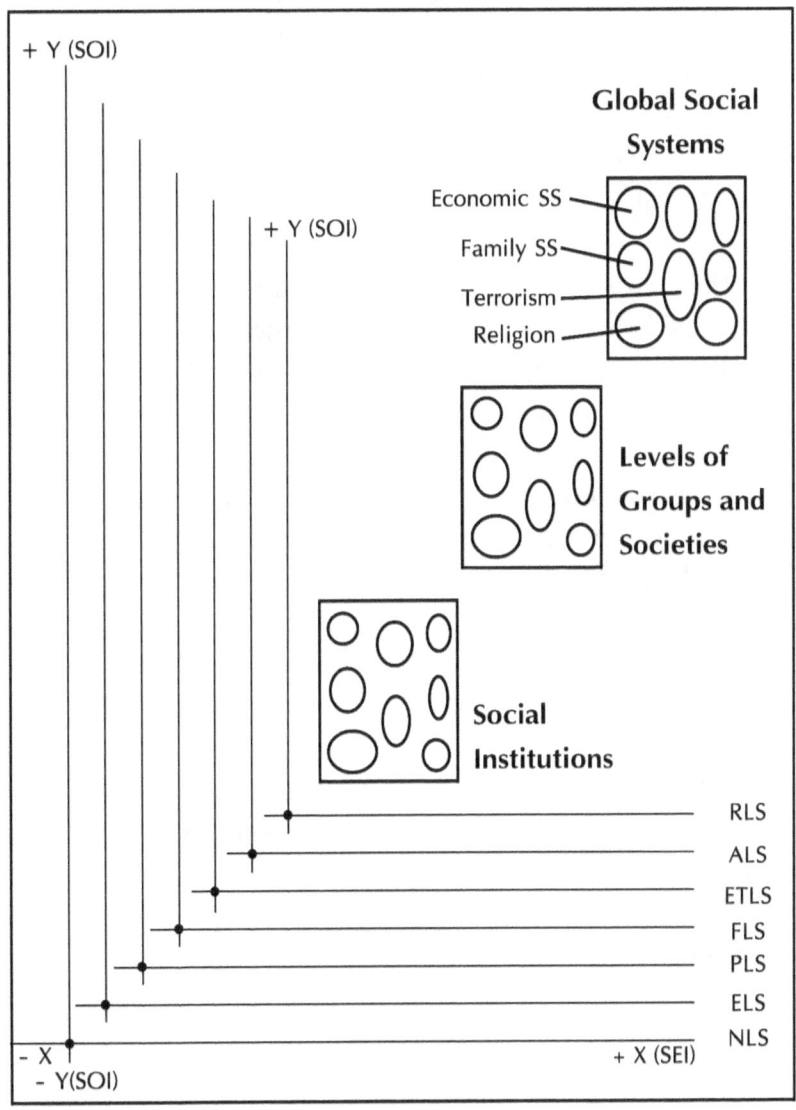

6.3 System Model of Knowledge

We may state that **social knowledge** is a phrase used here to refer to the knowledge about social phenomena. The spectrum of social knowledge must be first divided into *scientific social knowledge* and *mystic social knowledge*, in accordance with our faculties of scientific mind and mystic mind respectively. Obviously, in the first kind, social phenomena are observed in physical terms; such knowledge is popularly called the *social science*.

It is now necessary to distinguish between Sociology and Normal Social Science. *We can explain that sociology is a people-oriented study; where as Normal Social Science is a role-oriented study.* [*]. The discipline of **Sociology** studies the social behaviors of individual human beings and groups participating in a particular social phenomenon. For that purpose, the norms, culture and roles of the individual members must be examined. Also the sociologist makes suitable assumptions about the goals of groups pertaining to the concerned society.

But certain social phenomena have global extension; here the society under study covers the entire human population upon earth. Such a macro society possesses an interesting structure, which must be understood in terms of more abstract concepts. This kind of scientific enquiry about *social structure* is called **Normal Social Science**. The global social phenomenon as a whole is treated as a network of roles, which appears as a thing (machine or process). Then the cause-effect relations and other relevant sensible properties of component roles are studied using scientific method

We can classify the scientific and mystic disciplines under **social knowledge** as per the following table.

Table : Classification of Factual Knowledge about Social Phenomena. [# 4][*]

Scientific Social Knowledge (Social Sciences)	*Normal Social Sciences* such as anthropology, economics, political science, family science and ethics
	Sociology
Mystic Social Knowledge	*Mystical facts* under religion and art.

There is considerable uniformity in the methodological schemes of diverse kinds of knowledge. The uniform method is denoted by TyHDTI. This involves the integration of deductive proposition (DP) and inductive proposition (IP), which is achieved by the system philosophy of mind. So we can propose that our mind is a system of *brain and other parts of nervous system (BNS)* and *consciousness* as per the X-Y model.

Adopting a layered view of mind, there are separate faculties of scientific mind and mystic mind for producing and knowing the concerned societies and social systems. In other words, these faculties exist as systems of BNS and consciousness; it produce the TyHDTI schemes pertaining to various kinds of knowledge like natural sciences, sociology, economics, political science, family science, ethics, religion and art respectively. The synthesis in this respect is shown by the following diagram called ***System model of knowledge***.

Chapter 6 / Diagram 2
Systems Model of Knowledge - Synthesis of Methodology of Natutal Science and Social Subjects

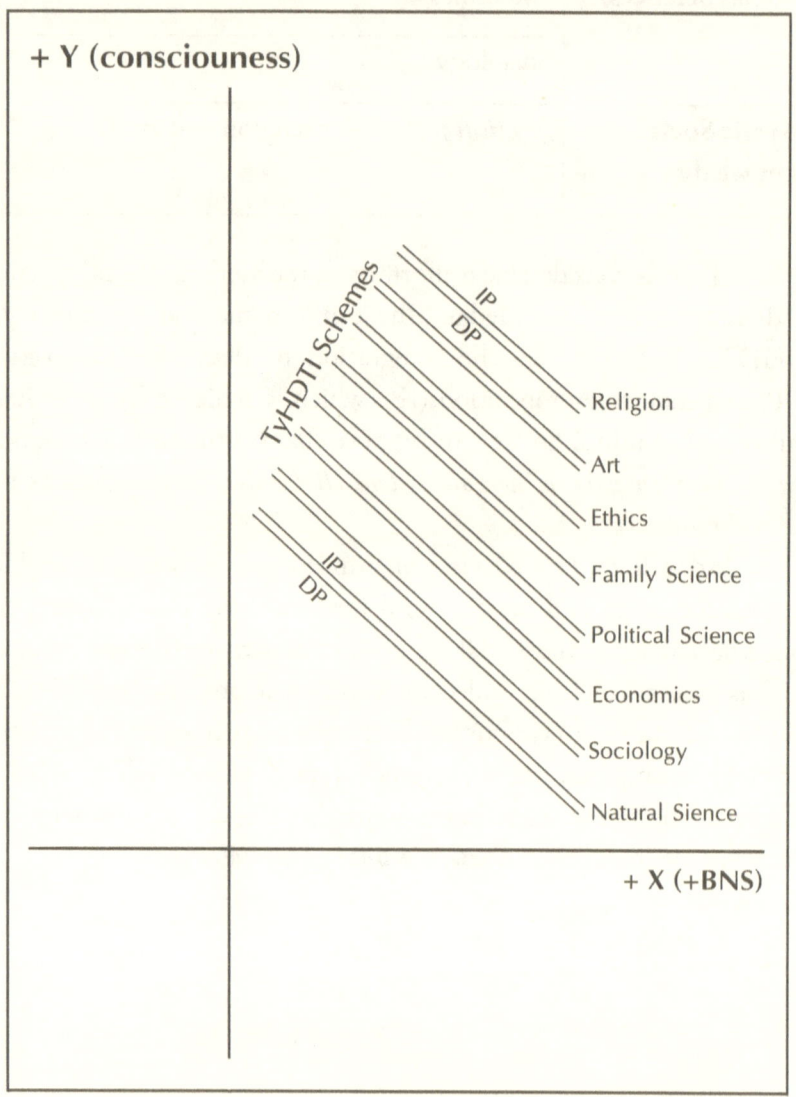

This diagram shows the existence of factual knowledge, through scientific and mystic methods, in the first quadrant of X-Y coordinate system. Note that there are two levels for a TyHDTI scheme for each discipline -- first is the level of DP consisting of propositions under Ty, H and D and secondly the level of IP consisting of proposition under T and I. Now we can link the diagram of knowledge to the most fundamental idea of six worldviews introduced in chapter 3, which are synthesized into a 2x2 table according to content view, process view, rationalism and empiricism. Philosophical knowledge is not shown in this diagram. But we can include Philosophy by drawing its DP and IP lines below that of natural science.

For example, in the case of natural science and social science, we undertake physicalisation of things. The scientific study is firstly carried out through content view adopting the worldview of materialism or naturalism. Secondly, scientist can adhere to physical process view also. These two world views promote the empirical method of study without recognizing the rational aspect of knowledge. But it has been explained earlier that empiricism cannot be sustained ipso facto. The right path is to achieve the synthesis of rationalism and empiricism, and it is the original idea of TyHDTI scheme underlying the *system model of knowledge.*

For the purpose of synthesis, each TyHDTI scheme under a particular discipline is split into two lines of DP and IP respectively as shown in the above diagram. However, the levels of deduction and induction work together in the process of producing meaningful laws and inferences about concerned phenomena. The X-axis denotes the empirical component and Y-axis represents the rational component pertaining to each proposition included in the lines of DP or IP.

As an example, the mathematical proposition *2 + 5 = 7* belongs to the class DP of natural science; it has the dual components of X and Y. In other words, the rational component and empirical component of scientific mind together function as a system for producing the said proportion. It is not reasonable to say that deductive proportion is rational only, while the inductive proportion is empirical exclusively. In this manner, the System Model of Knowledge resolves the conflict between rationalism and empiricism.

Further, we may reiterate that the theory of a particular kind of knowledge must describe the salient aspects of existence of the concerned *life systems,* which cause the multiple levels of social phenomena. This is the *system theory of justification of knowledge,* which is linked to methodology and source. The treatise about a comprehensive theory of truth is reserved for the final chapter of this book. In that occasion, we will expound that the notion of *good and bad social systems* enables us to link fact with value.

A brief summary

Throughout the foregoing discussion, our concern has been to find the laws about social world that has a hierarchical order from groups to societies. There are numerous social institutions which preexist for imposing the values and goals to social relationships. Tracing the origin of institution to the dawn of human civilization, we have introduced the innovative idea of *seven life systems.* There are non-overlapping institutions at the global level. They exist by the dualistic relation between SEI and SOI pertaining to various faculties of human mind. We can aggregate the societies of a particular class to a social system at the global level on the basis of the characteristic purposes. The spectrum of numerous social systems can be analyzed philosophically by resorting to the notion of *seven life systems.*

The seven life systems exist depending on the structure of human mind. Consequently the *system philosophy of mind and ultimate reality* provides the *final justification* for the laws and inferences of knowledge under various disciplines. So it will pave the way for the epistemological synthesis of the entire spectrum of human knowledge.

NOTES of Chapter 6

1. **The main reference books** used for this chapter are the following:

Alex Inkeles (1993); Francis Abraham (1993); Hicks, John (1979); Levin, William C (1984); Lipsey, Richard (1983); Lucy Mair (1992); Martin Hollis (2000); Michael Haralambos and Robin Heald (1990).

2   Levin, William C (1984), *Sociological Ideas,* pages 218 - 220

3   The theory about *seven life systems* effectively links factual knowledge about social system with value. This theory is originally conceived due to the influence of the famous book *On Ethics and Economics* (1990) published by Amartya Sen.

4   The definition of Normal Social Sciences and its distinction from sociology, as presented here, is my original idea.

Chapter 7

Comprehensive View about Truth

****

7.1 The Dilemmas about Scientific Truth

7.2 Issues about Religious Truth

7.3 System Philosophy of Truth

Author's main original ideas are marked by [*].

The mark [#] gives the number of note at the end.

****

The present aim is to solve the dilemmas associated with the notion of truth as introduced in the second chapter. Based on the formal definition of knowledge as a *justified true belief*, we have already covered the epistemological issues pertaining to methodology, source and justification of propositions under physical and biological sciences as well as theory of reality. And, it has been indicated that the analysis can be extended to other kinds of knowledge such as religion and art.

Accordingly, truth is an additional condition of a justified belief for becoming knowledge. We can adopt the definition: **Truth is the property of a justified belief that it corresponds to an actual *state of affairs of the universe*.** It may be recalled that the phrase *state of affairs* denotes a thing or event actually existing in this universe, irrespective of whether any particular person is aware of it. Accordingly, truth is the relation between a proposition and an object in the universe. Here lies the philosophical dilemma.

Our knowledge (proposition) about a thing is solely based on our sense experiences occurring in our mind. It is expedient to treat mind as similar to the screen in a movie theater; images about external object appear in mind just like the characters playing their roles on the screen. The knowing self is analogous to the spectator of the happenings on the screen. Hence mind functions as a screen placed between self and external object. We cannot go beyond the screen of our mental states including images, feelings and ideas. Since we are unable to access the external object, how can we judge that our propositions correspond to the actual *state of affairs* of the universe? This issue makes truth as the most difficult topic of philosophy requiring great intellectual attention. [# 1][*]

The controversy about truth arises in the first place due to the fact that there are two kinds of justification – infallible and fallible – respectively pertaining to deductive propositions and inductive propositions. These two kinds of propositions are combined in the systematic way of TyHDTI scheme to produce all kinds of knowledge grouped under science, religion, art and other fields.

Deductive propositions are normally classified into Theory, Hypothesis and Deduction. These are connected by the logical method of syllogism that is infallible; hence deductive propositions are said to possess *necessary truth*. On the other hand, inductive propositions pertain to Testing and Induction, where the evidences cannot be absolutely certain. Such propositions can have only fallible justification, because it suffers from the problem of induction. In this situation, the truth of inductive propositions is qualified as *contingent truth*, because it depends on the validity of evidences collected so far. Further, in any discipline, there are different shades of necessary truth and contingent truth, in accordance with various *worldviews* adopted. Thus, in the case of science for example, we can differentiate between the TyHDTI schemes under mechanistic worldview and physical process worldview.

Similar study of truth is possible with respect to religion and other subjects. More specifically, in contrast to scientific truth, it is necessary to consider the truth applicable to propositions in religion and art also. Then many difficulties of a different order would emerge since the propositions pertaining to religious faith and aesthetic experience consist of mystical ideas involving symbolic and metaphoric use of language. The states of affairs constituting religious world and artistic world belong to our social world, as compared to the natural world pertaining to science. Obviously the nonscientific propositions call for alternative criterion of truth. Since the meaning of religious words goes beyond the literal and ordinary sense, we need a higher theory of truth.

In this context we must discuss the question: how can we account for the different kinds of truth in science, religion and art? Further, based on the classification of knowledge into fact and value, we can talk about the corresponding levels of truth as factual truth and value truth. **More importantly, we desire to find a method for unifying the truths of various branches of knowledge.**

In the course of this study, it is necessary to examine the popular view that *truth* is intrinsically related to the notion of *good*, which is central to ethics. Can we deduce that truth and goodness are essentially the same? The question like what is good and what is bad would enlighten us to reach the definition of truth in tune with the comprehensive vision of reality. In this pursuit, the tenets of System philosophy are expected to help us in evolving a unified definition of truth by linking the various kinds of factual truth to the property of value. [# 2][*]

7.1 The Dilemmas about Scientific Truth

The guiding spirit for the ensuing discussion is the possibility that a *justified belief* may be false. For examples, consider the propositions "life exists only on earth" and "there is no living being in other planets and stars". These beliefs are justified now, but it can be false if new evidence come about life in any place other than earth.

As the final stage in philosophy of science, we have to deliberate upon the criteria of truth pertaining to deductive propositions and inductive propositions of scientific disciplines. A factual proposition can be treated as knowledge only if it is true; so we come to the question: What is it for such a proposition to be true? Here we have to take into account the distinction between content view and process view. Initially we are interested in the content view with respect to the two kinds of scientific propositions – deductive propositions and inductive propositions – for which we expect correspondingly different types of truth. The process view of scientific truth will be discussed afterwards. [# 3].

Truth of Deductive Propositions under Content View

Our mind has creative ability to combine logical thinking and past experience for generating certain general principles such as definitions, mathematical ideas and fundamental properties of things. In the context of scientific knowledge, the theory is composed of such general principles. One general principle can serve as the premise of an argument. Using a premise and particular factual information we can logically deduce another statement called conclusion – this process is termed as *method of deduction*. It was Aristotle, the Greek Philosopher, who first explained that deductive method consists of a *syllogism* of premise, fact and conclusion; it can be illustrated by the following example.

A - Shops are closed on holidays (premise)
B - Coming Tuesday is a holiday (fact)
C - Therefore, shops will be closed on coming Tuesday (conclusion)

An important feature of the syllogism is that the truth of conclusion is contained in the premise. This kind of truth is called **necessary truth** – *it is the same as the **coherence theory** to be explained below*. In other words, the conclusion is necessarily true due to the rational process irrespective of whether the premise is true or false actually. The method of syllogism can be applied in various situations, including mathematics and religion also, provided the premise is an abstract idea or deductive proposition.

In this context it is worth noting that syllogism does not talk about the existence of anything in the world because it is identical to the conditional form "if the premise is true and a suitable fact is given, then the conclusion is true". We can convert the above example into the form "if shops are closed on holidays (premise) and coming Tuesday is a holiday (fact), then shops will be closed on coming Tuesday (conclusion)". Obviously this complex sentence does not say that shops and Tuesday exist in this universe. Similarly the mathematical sentence *3+18 = 21* can be expressed as a syllogism; but it does not talk about the existence of numbers denoting things like mangoes, cats, men and houses. We can construct syllogisms about God, soul, angel, heaven, hell and other supernatural entities for obtaining conclusions under religious faith, without implying that such entities have existence.

In this situation, it is important to explain that *necessary truth does not mean factual truth*. This implies that, even if the premise is factually false, we can get the *necessary truth of* the conclusion through valid syllogism; in that case the conclusion also would be false factually. The following example will illustrate this point.

A - All birds can fly.
B - Emu is a bird.
C - Therefore, Emu can fly.

We know that the premise is false, since there are birds like Emu and Ostrich which cannot fly. Hence the conclusion also is false. But, in a syllogism, the conclusion has necessary truth because it logically follows from the premise. This is what we mean by saying that the truth of conclusion is contained in the premise. We can argue that, since necessary truth does not imply factual truth, the syllogism is not a valid knowledge about the things in this world.

Additionally, it may be noted that the truth of deductive propositions – definitions, abstract concepts and logical deductions – is different from that of inductive propositions supported by direct sensory data. Then it is imperative to find a suitable criterion of truth for deductive propositions. When we consider value truth which pertains to the propositions about values, the connection with sensory experiences is very thin; the notion of value is mostly a rational concept. Hence, in all these cases of deductive propositions a theory of truth called **coherence theory** is proposed. Accordingly, necessary truth is constituted by the proper relation between propositions themselves.

Truth of Inductive Proposition under Content View

We may now confine our treatment to the inductive propositions of science. Typically, an inductive proposition is an inference about cause-effect derived out of experimental observations, collection of data and analysis. However, the scientist has to rely upon specific abstract concepts like definitions, fundamental properties of physical world and mathematical principles. Accordingly we can note that inductive inference is the result of an *experimental syllogism* with three parts, namely, abstract concepts (D) - fact (E) - inference (F). This syllogism is formulated within the framework of scientific method. The philosophical discussion pertaining to the inductive propositions deals with the salient features of the three stages D, E and F. In this situation, the question of truth is very complex due to the *problem of induction,* which has been elaborated earlier in the context of the empiricism of David Hume.

Inductive method consists of making a general inference about population based on some regularity observed in a sample. In technical language we infer that *all Js are Ks* based on the sample observation that *some Js are Ks*. It is clear that the inference from *some* to *all* has an element of probability. There is a chance for the regularity of sample to be invalid for the population as a whole. There is no logical necessity in the inductive inference since it is based on a sample. This issue is called the **problem of induction**. It essentially means the possibility that some future observation may refute the inference. The experiment for deriving an inductive inference (scientific law) gives only a sample of observations. So it is possible that some future experiment may contradict the inference or law.

Since the problem of induction cannot be removed from experimental inference, the notion of **contingent truth** is adopted. This means that the truth of inductive inference is contingent on sample observations. In other words, a particular inference is true so long as it is not falsified by a future experiment. Considering the possibility of refuting the inductive proposition (scientific law) by means of counter evidence in future, the criterion of truth is ambiguous here. However the underlying principle applied here is that the proposition or law represents the state of affairs of world.

The best theory of contingent truth pertaining to inductive propositions under content view is called **correspondence theory**. It holds that an inductive proposition is true if it corresponds to the actual state of affairs; this correspondence is to be established by experimental evidences. It may be added that this theory is normally used for deciding the truth of propositions about things, events and cause-effect relations known through our sense organs.

The critical analysis of *correspondence theory* reveals that there are many ambiguities in determining the correspondence between the proposition and the external object. To illustrate this point consider the proposition P: *this bridge is safe for vehicles*. The proposition P is constructed on the basis of scientific theories about the strength of bridge as well as suitable experiments. Being scientific knowledge, it is true only for the experiments conducted. We cannot say that P is true for all sorts of vehicles using the bridge in future. It is a fact that we do not have the capacity to anticipate the future events with a good degree of certainty. In this situation the practical application of correspondence theory has limited scope only. In other words, the truth of P depends on a set of unknown conditions of future. This statement challenges our absolute idea of truth as an immutable property. **We have to admit that truth varies according to change of circumstances.**

The *problem of truth* in classical science comes up due to the difference in the methodologies of rationalism and empiricism. It has been explained already that all deductive propositions are obtained through reasoning or logical thinking out of general principles. According to rationalism, our mind has creative ability to form general principles from past experience and it serves as the source for a variety of propositions such as definitions and mathematical ideas, which have *necessary truth*. On the other hand, as per empiricism, physical laws are inductive inferences based on past observations. Scientists do not have absolute logic or certainty for predicting future events. Taking into account of this fact, we can say that a scientific law has **contingent truth** only; it means that the truth of a particular law is dependent exclusively on the evidences from the experiments already conducted. Since scientific laws are likely to be revised in future, an absolute definition of truth is impossible.

When deductive propositions (Ty, H, D) and inductive propositions (T, I) have different kinds of truth, is there any method to synthesize deduction and induction? This is the unsolved problem regarding the notion of truth in classical science. We will address this question by extending our focus to the process view pertaining to modern science.

Physical Process View of Truth

What is the *theory of truth* related to **process view of science**, as exemplified in quantum mechanics, quantum cosmology, evolutionary biology and related disciplines? Here knowledge is defined as an activity of giving meaning to theoretical entities by collecting experimental evidences so as to eliminate the dichotomy between deductive and inductive propositions. In this situation, we have to consider the *pragmatic theory of truth* as the process version of contingent truth.

The propositions under process view would describe the context or circumstances causing the change in a phenomenon. It is a descriptive type of knowledge about activity, rather than about a thing. In this case we don't consider the phenomenon as such; hence the knowledge about context cannot be expressed in content view. Accordingly, correspondence theory is not valid here. Since process view is concerned with the aspects of change in the state of affairs of world, *it is impossible to say whether the changes exist actually*. Hence propositions about process or change, constructed through TyHDTI scheme, require a separate criterion of truth, which is popularly termed as **pragmatic theory**. It is originally expounded by William James (1842-1910) and his followers in the context of the study of survival instincts of organisms. Accordingly, a proposition is taken as true if the concerned knower derives practical benefit from that knowledge. In the case of science, true propositions satisfy certain practical purposes of our ordinary life.

In the second half of twentieth century, the pragmatic theory of truth became the foundation of *post modernism*, which is defined here as the approach to knowledge that takes into account the changing aspects of language, culture, gender, tastes, etc. This method naturally leads to the position of *relativism* with regard to truth. The correspondence of a proposition with the actual state of affairs of world must be conceived in different social contexts, causing relative notions of truth. Accordingly, there is no single truth, but a lot of competing truths is possible. [# 4].

Now we shall explain that, from ontological considerations, the physical process view of truth must lean on the content view of truth considered earlier. It must be admitted that permanence and change are complementary aspects of a thing. For example, when we say that climate is changing, we are talking about the phenomenon called climate. We cannot define climate unless it exists with certain static features; it means that climate has permanence at least for a moment. Such definitions related to permanence are conceived in content view. Something must exist so that it can undergo change. *So the truth of propositions under physical process view depends on the truth of concerned propositions under content view.* Thus we are back to the conflict between rationalism and empiricism leading to the dichotomy of necessary truth and contingent truth. [# 5][*].

7.2 Issues about Religious Truth

It is a difficult matter to judge whether religious entities like God, soul, heaven and hell are actual state-of-affairs of the world. In order to make this point clear, let us compare a scientific proposition A with a religious proposition B.

A: This is a building used for prayers.
B: This is a church with the presence of God.

Obviously A represents the actual state-of-affairs of world, which can be ascertained using our sense experience and scientific thinking. On the other hand, B expresses metaphoric language connected with religious faith and hence it is not about the actual state-of-affairs of world. We hold that the terms 'physical' and 'natural' pertain to the actual state-of-affairs of world.

In this situation, the truth of propositions under religion presents many difficulties of a different order. Since the meaning of religious words goes beyond the literal and ordinary sense, we need another kind of theory of truth, as compared to scientific truth. This philosophical point will be elaborated in the next section for finding a method for unification.

7.3 System Philosophy of Truth [# 6][*]

The concept of truth is the overarching principle in the entire corpus of philosophy and especially in the epistemology of various kinds of knowledge. But the various dimensions of truth as mentioned above refute the possibility of an absolute idea of truth. The reason may be stated here. Till now we presented the famous definition that **knowledge is justified true belief (JTB)** at the level of individual propositions, which are simple and complex. The riddles of epistemology – with the parts called methodology, source, justification and truth -- spring from this practice of studying the topic from the level of individual propositions.

We can note that such issues remain unsolved due to the lack of integrative view regarding the parts of knowledge. ***In this situation, for clearing the age-old controversies, I have proposed in the first chapter itself to analyze knowledge at the level of laws produced by TyHDTI schemes and also at the higher levels of concerned disciplines. This approach is to be applied for the deliberation about truth also.*** It can be realized that simple facts as well as propositions of direct perception are elementary forms of TyHDTI scheme, because they involve both deduction of abstract concepts as well as induction from sense experience.

From the foregoing, we realize that there are different criteria of truth with respect to deductive and inductive propositions. As a result, there raises the question: Which type of proposition is true for the particular discipline? Generally, how can we compare the criteria of truth applicable to various kinds of human knowledge? A comprehensive understanding of the nature of our mind as well as the theories of justification is required for constructing the correct theory of truth. In that pursuit, most basically we need the notion of *worldview* with sufficient details, which would lead to the theory to be employed in the *TyHDTI schemes* of a discipline. We have already encountered with six different kinds of worldviews which account for the corresponding theories of truth.

For a clear perspective on **scientific truth**, we must consider the solution of the outstanding issues pertaining to the ontology, methodology, source, and justification related to scientific propositions. In this pursuit, as established in the previous chapters of this book, we have to resort to the tenets of System Philosophy. Our concluding insights in this regard can be arranged as following for the brevity of presentation.

In the context of scientific study of universe, it is expedient to divide it into three levels – inanimate world, living world and social world – that are observed in physical terms. We will first take up the enquiry about the truth of scientific laws, which pertain to the physical version of inanimate world.

Through our innovative philosophy we have shown that scientific laws are formulated through TyHDTI scheme that involves the judicious combination of deductive propositions and inductive propositions with respect to the phenomenon under study. The confusion about truth arises from the traditional view that deductive proposition has *necessary truth* while inductive proposition holds *contingent truth*. The point of contention is about the combination of these two types of truths in a systematic manner.

The basic part of deductive propositions is the theory about the constitution of physical world. It has been explained earlier that the theory deals with cosmological entities which are presumed to exist as per the position of scientific realism. The existence of such theoretical entities is accepted in the form of deductive inference known as *Inference to the Best Explanation (IBE)*. Proposing that quarks, strings, membranes and allied mysterious objects have existence is the best way to explain the development of physical world and its laws. The progress of science is attributed to the discovery of deeper and more general entities of theory *by way of deduction*, to be proved subsequently through external evidences. Note that empiricist philosophers wrongly held that theory or IBE is proposed in the course of inductive reasoning.

In this situation, theoretical propositions are usually said to be rational (or deductive) propositions as they involve the creative abstraction performed by our mind. But the concrete things pertaining to the theoretical entities are experienced using our sense organs and it results in the inductive propositions or empirical knowledge. Hence we need the synthesis of rationalism and empiricism. According to the System Philosophy of Mind, the so called deductive proposition is in fact a combination of rational ideas and empirical ideas, *having higher proportion of rational part*. Keeping this principle in view, we can say in an innovative way that *IBE about theoretical entities as well as other deductive propositions have **necessary truth*** as it is based on the rational syllogism. This is a modification to the empirical argument that necessary truth is applicable only to analytic statements appearing in mathematics as well as logic and hence not relevant in the context of scientific theory.

Coming to the truth of inductive propositions pertaining to the stages of Testing and Inductive inference, the problem of induction is the persisting cause of ambiguity. As the scientific law is an inference based on finite number of observations, there is no guarantee that it will not be falsified by future observations. Empirical philosophers from David Hume onwards denied the creative power of our scientific mind; they treated theoretical concepts (deductive propositions) as abbreviations of experimental observations. In this situation, empiricists have stumbled upon the problem of induction so as to suggest that there is no truth in science. In the light of System Philosophy of Mind, we can conceive that the inductive propositions are combinations of rational ideas and empirical ideas, *having higher proportion of empirical part*. In this way, the **contingent truth** of inductive method has both rational and empirical aspects, though it is conventionally treated as empirical.

Since scientific knowledge is formed by TyHDTI scheme, it rests on the twin pillars of deduction and induction. So the foregoing ideas suggest that the truth of a scientific law is a combination of necessity and contingency. For example, consider the scientific laws about electricity. Such laws are true if:

(1) Electricity exists with properties revealed through suitable experiments. These facts have *contingent truth*.
(2) Electricity is caused by the movement of electrons in the conductor as well as by other subatomic phenomena. The existence of these underlying aspects is known through rational or deductive propositions included in the theory of electricity. It is ultimately based on the theory of matter, which has *necessary truth* as explained above.

The existence of electricity as a state of affairs of physical world can be accepted as *true* because theory and experiment are mutually supportive. The foregoing analysis indicates that necessary truth and contingent truth would together form a system of truth. This point leads to the following principle: *Scientific truth is a system of necessary truth of theory and contingent truth of inductive propositions.* At the same time, these kinds of truth are two levels of rational ideas and empirical ideas in different proportions. *Hence we suggest the phrase **Necessary-Contingent Truth (NCT)** to refer to the duality inherent in scientific truth.* The essential aspects of this notion of truth are presented below [# 7] [*].

- ❖ Scientific laws possess NCT because they conform to the principles of ontology, methodology, source, and justification. However it suffers from the problem of induction since future is uncertain in the case of experimental method. Additionally, the possibility of underdetermination worsens the ambiguity of scientific truth. We can hold that there is no absolute truth in science. Scientific truth is of pragmatic or practical nature, though based on appropriate methodology and evidences.
- ❖ The explanatory power as well as predictive efficiency of scientific law mainly depends on the accuracy of theory. Corresponding to the deepening levels of physical world, there is a hierarchy of theories in the increasing order of generality; it accounts for the success of scientific enterprise. The pyramidal architecture of theories implies that the theories are internally consistent, and mutually supporting.
- ❖ Each theory is produced by our mental faculty, which has rational part (Consciousness) and empirical part (Brain and other parts of Nervous System, denoted by BNS). So the X-Y model can be gainfully used for depicting the levels of theories, in conformity with the system model of universe. Theory has the aspects of logic and other forms of rationality; and it is substantiated by the experimental data.
- ❖ The theoretical structure of science ensures that its laws are reliable for practical purpose. For example, the theories of astrophysics hold well to the statement that sun will rise tomorrow. To put it simply, there is 99% certainty about the occurrence of tomorrow so that we can plan our activities.

However, the scientific view does not rule out the rarest event of collapse of solar system. Similarly, we can accept the proposition "there is no living being in astronomical bodies other than earth" – it is true for all practical purposes, though it has no certainty.

❖ The vital reason for the success of science is that the objects of nature exist as systems of body and mind; the physical world is the scientific reduction of such systems. The concerned theory of reality has been presented in fifth chapter. The natural laws are manifestations of ultimate reality which are converted into physical laws through scientific method. In other words, the phenomenal stability of physical laws is derived from the constitution of nature that appears as a system with mysterious features.

To sum up the above discussion, scientific truth is defined here as ***necessary-contingent truth (NCT)*** that is supported by the epistemological factors called methodology, source, and justification. Moreover it synthesizes the opposite aspects of coherence theory (necessary truth) and correspondence theory (contingent truth). The integrative model of truth is amenable to process view also, when we study the changing aspects of physical world. Scientific knowledge under process view is a combination of deduction and induction with respect to changing phenomena. Here we resort to the idea that knowledge is a manufacturing process involving creativity and sensory experience; in this situation, it does not have the baggage of pragmatism or instrumentalism.

We have developed the foregoing model of truth in the context of physical science pertaining to inanimate things. Next we may focus on the biological world. Though biological science and psychology also possess NCT, it is in decreasing order of preciseness or rigor as compared to that of physical science. Here *preciseness* means greater probability of remaining true in future experiments. The experimental methods pertaining to these disciplines do not yield clear cut laws of cause-effect. So we can say that factual propositions pertaining to the three levels of natural science -- physical science, biology and psychology -- have separate levels of NCT in decreasing order of preciseness.

Religious Truth

Coming to the level of social world, we are concerned with the social perspective of knowledge and its truth. It is to be recognized that natural sciences are the result of human action - research, experiments, inferences, etc. - in social context. As explained already, scientific knowledge is produced when individuals participate in NLS with the objective of viewing natural objects as physical things. Similarly, there are six man-made life systems -- ELS, PLS, FLS, ETLS, RLS and ALS -- as defined earlier. We get separate kinds of factual knowledge consisting of deductive and inductive propositions with regard to each of these life systems.

The most important topic to be considered here is the truth of religious propositions. It involves the extension from the treatment of religious justification given in previous book. We may recall that the various propositions of faith are constructed by our mystic (religious) mind on the basis of the religious concepts about supernatural entities like God, soul, sin, heaven and hell. Such concepts constitute the theory of Religious Life System (RLS). Hence the existence of RLS is the **justification** for the mystical propositions about God and other aspects of religious worship. It is emphasized that RLS is organized by the faculty of mystic mind, which is complementary to intellectual mind, of human beings.

In order to formulate the propositions like "God exists" and "God is love", a believer depends on the theory about RLS and also the evidences of mystic (religious) experiences. Through our innovative philosophy we have shown that such inferences or laws are formulated through TyHDTI scheme that involves the judicious combination of deductive propositions and inductive propositions with respect to the phenomenon under study. The confusion about truth arises from the traditional view that deductive proposition has necessary truth while inductive proposition holds contingent truth. Here the point of contention is about the combination of these two types of truths in a coherent manner. Happily now we have a technique to solve such issues and arrive at the **comprehensive definition of truth** as applicable to the mystic knowledge pertaining to religion.

Science and religion can be treated as two technologies for producing knowledge in the factory called human mind. By this logic we can hold that religious knowledge has *necessary-contingent truth (NCT)*, based on the validity of method, source and justification in our mystic mind. The same principle applies to the case of spiritual science also, including mystical process science and parapsychology. This innovative theory of truth is capable of resolving the dichotomy between rationalism and empiricism in the propositions about supernatural things conceived through mystic mind.

> *A few additional insights are in order. The foregoing notion of NCT implies that the truth of a particular piece of knowledge is a systematic conjecture, because it is derived from the framework of theory and experiment. The worldview is the foundation of the procedure of constructing knowledge; truth is the highest point in that project. In other words, truth is our judgment; it is not external to our mind. So we have to avoid realism in the case of truth also.*

Final Unification

Generally considering the existence of seven life systems, we can apply the system model of knowledge and truth to the propositions of social sciences as well as to the mystical facts of religion and art. But the objectivity of NCT has a decreasing order for these disciplines. *This relative notion of factual truth is troublesome, because we are used to form the concept of truth in absolute manner.* We need the **synthesis of different levels of factual truths** so as to answer the following questions:

- Why are there different levels of *necessary-contingent truth?*
- *How can we unify such levels of* factual truths?
- *What is the definition of truth as applicable to all fields of* knowledge?

Here we require the ontological definition of truth, which can be applied uniformly to natural sciences, social sciences and mystical facts. As long as we consider truth as fact, there are many kinds of *necessary-contingent truth*, which stand separately in pluralistic manner. For comparing and unifying two entities, we must use the notion of value or purpose. **In the unifying definition of truth, we must treat truth as a value, which is in contrast to falsehood.** Accordingly, truth and falsehood are opposite values allied to the ethical norms of good and bad. Recall the inconsistencies in the thoughts of illustrious philosophers of past, when they confronted the notions of good and truth. The hitherto history of philosophy has suffered from the lack of a suitable method for linking opposite values of good and bad ontologically.[# 8].

We may assert that values are expressions of our purposes, which serve as the causes of our actions. Such values or purposes have opposite directions of good and bad. This principle can be applied to the life systems like NLS, ELS and so on, which are formed by respective class of values. We have already proposed that the values constituting a life system have the components of self-interest (SEI) and society-interest (SOI). When SEI is represented by X-axis, its positive side shows good aspect, while the negative side shows bad aspect. Similarly SOI is represented by Y-axis, with positive and negative sides. Hence in the X-Y model of a life system, the first quadrant shows the good systems, while the third quadrant shows the bad systems. We can call this as the **System Model of Good and Bad** in social systems.

Now let us consider the good systems and bad systems appearing in the first and third quadrants respectively of the above model of ethics. This gives us the criterion of truth in contrast to falsehood. The levels of NCT pertaining to the knowledge about good systems have the property of truth. Accordingly, **truth is defined as the *necessary-contingent***

***truth* pertaining to the knowledge about good systems. Conversely, falsehood is defined as the *necessary-contingent truth* pertaining to the knowledge about bad systems.** The various disciplines of natural sciences, social sciences and mystical facts have the properties of truth and falsehood depending on the good and bad aspects of the systems formed by social relations. In this manner **we can link fact with value.** Truth and falsehood exist ontologically as per the system model of self-interest (SEI) and society-interest (SOI). This conception of truth is called the **System Model of Truth,** which can be illustrated by the diagram of X-Y coordinate system.

A simple example may be given for clarifying the above *system model of truth*. Mathematics is normally treated as a value-neutral subject; it is included in the class of deductive propositions pertaining to NCT. However, the foregoing analysis enables us to link mathematics to value. If mathematics is used in the production of knowledge about good systems – for example, medicines, nutritive food and so on – then we can treat that mathematics have truth. On the reverse side, we see that mathematics has falsehood if it is used for the knowledge about atom bomb and chemical poison.

The advantage of the *system model of truth* is that it solves the age-old problems of ethics. In this context, we may examine the meaning of the phrase "normative approach" used widely in the literature of social sciences. When we want to make a judgment as to *what ought to be* or *what should be*, then we are adopting the normative approach. This involves the application of ethical principle to the concerned factual knowledge. Evidently normative approach is categorically different from factual approach.

Let us take an example from economics. When we study the conditions of increasing employment in a country, it is a kind of fact. But normative approach seeks to answer the question: what should be the level of employment to be aimed at in a particular year? This normative question contains a sense of duty or commitment. Here we require a criterion of *good* or *desired goal*. The crux of the issue in this situation is to define good and bad in an abstract way so that it can be used in practical cases. In other words, we must be able *to link fact with* *value* for arriving at the definition of good and bad; it is the central problem in the discipline of ethics and economics. [# 9][*].

When we take up the study of ethical life system (ETLS) in a later occasion, we will specifically consider the normative approach of ethical question pertaining to economics and other social sciences. In anticipation of the *system ethics* to be developed for that purpose, we can say that good and bad are defined in terms of the positive and negative sides of the system model of ETLS in which self-interest (SEI) is the X-axis and society-interest (SOI) is the Y-axis. This norm of ethics is based on the goals of our social behavior. The systematic analysis of various social sciences as well as the formulation of system ethics is reserved for a separate volume.

Summary of System Philosophy of Truth:

- ❖ We can adopt the definition: *truth is the property of a justified belief that it corresponds to an actual state of affairs of the universe.* Here the philosophical issue is -- how can we reconcile between the various kinds of truth in science, religion, art and other fields of knowledge?
- ❖ Four propositions are cited now to hint at the issue of truth: (a) 2 + 14 = 16, (b) life exists only on earth, (c) this bridge is safe for vehicles, and (d) God is love.
- ❖ The traditional theories of truth are proposed at the level of individual propositions under the two categories – deductive propositions and inductive propositions. These categories are said to have *necessary truth* and *contingent truth* respectively. The related theories of truth are coherence theory and correspondence theory, which involve the conflict between rationalism and empiricism.
- ❖ We can note that truth is not an absolute idea. The meaning of truth varies in accordance with the methodology and justification of various kinds of knowledge. How can we explain this view? Is it possible to get a unified notion of truth?

- **The innovative idea of this chapter** is that truth must be discussed at the level of TyHDTI schemes, rather than for individual propositions. Since every TyHDTI scheme is a combination of deductive propositions and inductive propositions, it is proposed here that truth exists as the system of *necessary truth* and *contingent truth*. In other words, truth of a TyHDTI scheme is a system formed by the necessary truth of theory and contingent truth of inductive propositions. This principle is expressed by the phrase **necessary-contingent truth (NCT).** For example, the justified scientific proposition "living beings exist only on earth" has NCT, since it is based on proper theory and sound evidences collected so far. However it suffers from the problem of induction since future is uncertain in the case of experimental method.
- The theoretical structure of science ensures that its laws are reliable for practical purpose. To put it simply, there is 99% certainty about the occurrence of tomorrow so that we can plan our activities. However, the scientific view does not rule out the rarest event of collapse of solar system. Similarly, we can accept the proposition "living beings exist only on earth"; it is true for all practical purposes, though without certainty.
- The physical world is the scientific reduction of the systems of body and mind, which constitute the nature. Hence, the phenomenal stability of physical laws is derived from the constitution of nature that appears as a giant system.
- There are different levels of NCT for science, religion and other fields. But all types of knowledge exist on the basis of the dual goals of human mind, namely self-interest (SEI) and society-interest (SOI). Using the X-Y coordinate model of SEI and SOI, we can define *good* and *bad* as complementary and opposite qualities. This principle serves as the norm for ethics. And, it is the method for linking fact with value. Thus we get the *System Model of Truth*, which shows the ontological existence of truth and falsehood.

NOTES of Chapter 7

1 Descartes originally proposed the principle that mind is like the screen of a theater. Daniel Dennett calls this model of mind as Cartesian Theater. The discussion about this view is available in Leahey (2005), *A History of Psychology – Main currents in Psychological Thought*, page137.

2 Here the list of main reference books is the same as that of first chapter. This section presents the criticism on the traditional theories of truth so as to lead to the innovative system model of truth. The originality of the treatise can be realized by comparing this presentation with the relevant topics in the books of reference. Specifically, comparison may be made with the discussion of correspondence theory and coherence theory in Chatterjee (1988), Ewing(1994) and Hospers (1997

3 The details of philosophy of science have been discussed in the earlier book *Origin of Universe*.

4 See Caputo (2013), *Truth – Philosophy in Transit*, page 7.

5 It may be mentioned that shifting from process view to content view is the unique project of System Philosophy in order to tackle the issue of existence.

6 It may be noted that the System Philosophy of Truth is developed from the entire corpus of ideas presented so far in this book.

7. These points would serve as the final answer to the problems of philosophy of science.

8 This issue is linked to the problem of evil discussed earlier.

#9 *On Ethics and Economics,* of Amartya Sen (1990), is an important book about evaluating economic behavior using ethical principles. However, the treatment of Sen is in the empirical perspective of welfare, without considering the basic philosophical problems of ethics.

Bibliography

Alan Grafen and Mark Ridley (2007). *Richard Dawkins How a Scientist Changed the Way We Think,* (Oxford University Press, Paperback Edition)

Alex Inkeles (1993), *What is Sociology? – An Introduction to the Discipline and Profession,* (Prentice Hall of India Limited, New Delhi, Tenth Indian Reprint, 1993)

Ammar Al Chalabi, Martin R. Turner and R. Shane Delamont (2007), *The Brain – A Beginner's Guide,* (One World- Oxford, England, First South Asian Edition, 2007)

Anthony Harrison– Barbet (1990), *Mastering Philosophy,* (Macmillan, London, 1990).

Armstrong, Karen (1998), *A History of God,* (Arrow Books, London, 1998)

Augustine Perumalil (2000), *Critical issues in the Philosophy of Science and Religion,* (Indian Institute of Science and Religion, Pune and ISPCK, Delhi, 2006).

Behe, Michael J. (1996), *Darwin's Black Box: The Biochemical Challenge to Evolution,* (New York: Touchstone Books, 1996).

Bird, Alexander (2003), *Philosophy of Science,* (Routledge, London, Indian Reprint, 2003).

Beiser, Arthur (2002), *Concepts of Modern Physics,* (Tata McGraw-Hill, New Delhi, Sixth Edition, Second Reprint, 2002)

Blackburn, Simon (1996), *The Oxford Dictionary of Philosophy,* (Oxford University Press, 1996)

Brennan, James F. (2005), *History and Systems of Psychology,* (Pearson education, Delhi, First Indian Reprint, 2005)

Brooke Noel Moore and Kenneth Bruder (2005), *Philosophy - The Power of Ideas,* (Tata McGraw-Hill Publishing Co. Ltd, New Delhi, sixth edition, 2005.)

Capra, Fritjof (1983), *The Turning Point,* (Flamingo, London, 1983).

Capra, Fritjof (1992), *The Tao of physics*, (Flamingo, London, Third edition, 1992)

Capra, Fritjof (1997), *The Web of Life*, (Flamingo, London, 1997).

Caputo, John D (2013), *Truth – Philosophy in Transit*, (Penguin books, 2013)

Chardin, Teilhard de (1965), *The Phenomenon of Man*, (Harper Torchbook Edition, New York, 1965)

Chatterjee, Margaret (1988), *Philosophical Enquiries*, (Motilal Banarsidas, Delhi, 1988)

Copleston, Frederick S.J. (1994), *A History of Philosophy, Vol. I – IX*, (Image Books, Doubleday, 1994).

Darwin, Charles (1859), *On The Origin of Species*, (Dover Edition, New York, 2006)

Davies, Brian (2000), *Philosophy of Religion*, (Oxford University Press, contains the reprint of Hume's article 'Of Miracles', 2000)

Davies, Paul (1995), *About Time – Einstein's Unfinished Revolution*, (Penguin Books, 1995)

Davies, Paul (2007), *Cosmic Jackpot – Why Our Universe is Just Right for Life*, (Houghton Mifflin Company New York, 2007).

Dawkins, Richard (1976), *The Selfish Gene*, (Oxford University Press, Oxford and New York, 1976)

Dawkins, Richard (2007), *The God Delusion*, (Black Swan, Transworld Publishers. London, 2007)

Dawkins, Richard (2009), *The Greatest Show On Earth: Evidence for Evolution*, (Bantam Press, 2009)

Dennett, Daniel (1991), *Consciousness Explained*, (Boston: Little, Brown, 1991).

Esposito, John, et al (2008), *Religion & Globalization: World Religions in Historical Perspective*, (Oxford University Press, New York, 2008)

Ewing A. C. (1994), *The Fundamental Questions of Philosophy*, (Allied Publishers Limited, New Delhi, 1994)

Feser, Edward (2006), *Philosophy of Mind*, (Oneworld, Oxford, 2006).

Francis Abraham (1993), *Modern Sociological Theory*, (Oxford University Press, New Delhi, Ninth Impression, 1993).

Frost. S. E (1989), *Basic Teachings of the Great Philosophers*, (Anchor Books, Doubleday, New York, 1942/1989)

George Thomas White Patrick **(1978)**, *Introduction to Philosophy,* (Surjeet publications, Delhi, 1978).

Grayling A.C. (Editor) (1995), *Philosophy: A Guide Through The Subject,* (Oxford University Press, London, 1995).

Grayling A.C. (Editor) (1998), *Philosophy 2 : Further Through The Subject,* (Oxford University Press, London, 1998).

Green, Brian (2005), *The Fabric of the Cosmos,* (Wintage Books, New York, 2005)

Gribbin, John (2008), *The Universe: A Biography,* (Penguin Books, London, 2008)

Griffin, David Ray (2000), *Religion and Scientific Naturalism,* (State University of New York, 2000)

Grolier Encyclopedia of Knowledge, Volumes 1 – 20, (Grolier Incorporated, Danbury, Connecticut, 1993).

Guttman, Burton (2007), *Evolution– A Beginner's Guide,* (One World- Oxford, England, First South Asian Edition, 2007)

Guttman, Griffiths, Suzuki and Cullis(2006), *Genetics – A Beginner's Guide,* (One World- Oxford, England, First South Asian Edition, 2006)

Guyer, Paul (2008), *Kant,* (Routledge, London and New York, 2006; first Indian Reprint, 2008)

Haught, John F. (2000), *God after Darwin – A Theology of Evolution,* (Westview Press, USA, 2000)

Haught, John F. (2001), *Responses to 101 Questions on God and Evolution,* (Paulist Press, USA, 2001)

Hawking, Stephen W. (1995), *A Brief History of Time,* (Bantam Books, 1995 edition)

Hawking, Stephen W. (2011), *The Grand Design,* (Bantam Books, 2011 edition)

Heil, John (2003), *Philosophy of Mind,* (Routledge, London and New York; Indian reprint, 2003),

Hicks, John (1979), *Causality in Economics,* (Basil Blackwell Oxford, Great Briton, 1979)

Hick, John H. (1994), *Philosophy of Religion,* (Prentice Hall of India Pvt. Ltd, fourth edition, 1994).

Hiriyanna M (1948), *The Essentials of Indian Philosophy,* (Motilal Banarsidas, Delhi, First Indian dition 1995)

Hospers, John (1997), *An Introduction to Philosophical Analysis,* (Allied Publishers Limited, Mumbai, 1997; Original publication by Prentice–Hall in 1953)

Ivor Leclerc (1958), *Whitehead's Metaphysics,* (George Allen and Unwin Ltd, London, 1958).

James.T.Shipman, Jerry. L. Adams and Jerry. D. Wilson (1987), *An Introduction to Physical Science,* (D.C. Heath and company 1987).

Jantsch, Erich (1989), *The Self-organizing Universe,* (Pergamon Press, 1989).

Job Kozhamthadam (editor) (2002), *Contemporary Science and Religion in Dialogue-Challenges and Opportunities,* (ASSR Publications, Jnana-Deepa Vidyapeeth, Pune, 2002),

Job Kozhamthadam (editor) (2003), *Science, Technology and Values,* (ASSR Publications, Jnana-Deepa Vidyapeeth, Pune, 2003),

Job Kozhamthadam (editor) (2004), *Religious Phenomena in a World of Science,* (ASSR Publications, Jnana-Deepa Vidyapeeth, Pune, 2004),

Job Kozhamthadam (editor) (2005), *Modern Science, Religion and The Quest for Unity,* (ASSR Publications, Jnana-Deepa Vidyapeeth, Pune, 2005),

Kant, Immanuel (2003), *Critique of Pure Reason,* translated by J. M.D. Meiklejohn, (Dover Publications, New York, 2003)

King, Richard (2000), *Indian Philosophy - An Introduction to Hindu and Buddhist Thought,* (Ane Books & Maya Publishers, New Delhi, 2000)

Kuhn, Thomas (1970), *The Structure of Scientific Revolutions,* (University of Chicago Press, 1970)

Kukla, Andre (1998), *Studies in Scientific Realism,* (Oxford University Press, 1998)

Ladyman, James (2002), *Understanding Philosophy of Science,* (Routledge, London, 2002)

Lavin.T. Z. (1989), *From Socrates to Sartre,* (Bantam Books, New York, 1989)

Leahey, Thomas Hardy (2005), *A History of Psychology – Main Currents in Psychological Thought,* (Pearson education, Delhi, first Indian reprint, 2005)

Levin, William C (1984), *Sociological Ideas,* (Wadsworth Publishing Company, California, 1984)

Lewens, Tim (2007), *Darwin,* (Routledge, London and New York, 2007)

Lipsey, Richard (1983), *An Introduction to Positive Economics,* (ELBS Edition, 1983).

Lucy Mair (1992). *An Introduction to Social Anthropology*, (Oxford University Press, New Delhi, Seventh Impression,1992)

Luke George (2004), *Saptaloka Darshanam- Samgraham*, (PGL Books, Changanachery, Kerala, 2004, in Malayalam language).

Luke, George (2015), *Jeevanum Parinamavum*, (PGL Books, Changanachery, Kerala, 2015, in Malayalam language).

Macquarrie, John (1985), *In Search of Deity – An Essay in Dialectical Theism*, (Cross road Publishing, New York, 1985)

Martin Curd and J. A. Cover (1998), *Philosophy of Science: The Central Issues*, (W. W. Norton & Company, New York / London,1998)

Martin Hollis (2000), *The Philosophy of Social Science*, (Cambridge University Press, First Indian Paperback Edition, 2000)

Masih. Y. (1995), *Introduction to Religious Philosophy*, (Motilal Banarsidas, Delhi, reprint 1995)

Maslin K.T. (2001), *An Introduction to the philosophy of Mind*, (polity Press, UK & USA, 2001)

Max Charlesworth (2006), *Philosophy and Religion- From Plato to Postmodernism*, (Oneworld, Oxford, First South Asian Edition, 2006)

Mayr, Ernst (1999), *This Is Biology*, (Universities Press India Limited, Hyderabad)

McGinn, Colin (1998), *The Character of Mind*, (Oxford University Press, 1998

Michael Haralambos and Robin Heald (1990), *Sociology - Themes and Perspectives*, (Oxford University Press, New Delhi, Tenth Impression,1990)

Michio Kaku and Jennifer Thompson (2007), *Beyond Einstein: The Cosmic Quest for the Theory of the Universe*, (Oxford University Press, New Delhi, 2007)

Miller, James B. (Editor) (2001), *An Evolving Dialogue - Theological and Scientific Perspectives on Evolution*, (Trinity Press International, Harrisburg, Pennsylvania, 2001)

Newton, Roger (2010), *The Truth of Science – Physical Theories and Reality*, (Viva books, New Delhi, reprint 2010)

O'Leary, Denyse (2004), *By Design or by Chance?*, (Augsburg Books, Minneapolis, 2004).

Panda N. C. (1999), *Maya in Physics*, (Motilal Banarsidas, Delhi, reprint 1999)

Parthasarathy A. (2000), *Vedanta Treatise*, (Vedanta Life Institute, Mumbai, 2000)

Pulingandla R (2005), *Fundamentals of Indian Philosophy*, (D. K. Printworld, New Delhi, Third Impression 2005)

Radhakrishnan S (1931), *Indian Philosophy*, volumes I and II (George Allen & Unwin, 1931)

Robert John Russell (Editor) (2004), *Fifty Years in Science and Religion – Ian G. Barbour and his Legacy*, (Ashgate Publishing Ltd, England and USA, 2004),

Robert John Russell, Nancey Murphy and C. J. Isham (Editors) (1999), *Quantum Cosmology and the Laws of Nature : Scientific Perspectives of Divine Action*, (Vatican Observatory Publications, Vatican City State and The Center for Theology and the Natural Sciences, Berkeley, California; second edition, 1999)

Romijn, Herms (2002), *Are Virtual Photons the Elementary Carriers of Consciousness?*, (Journal of Consciousness Studies, 9, No.1, 2002, pp 61-81).

Rosenberg, Alex (2000), *Philosophy of Science, a Contemporary Introduction*, (Routledge, London and New York, 2000),

Russell, Bertrand (1992), *The Problems of Philosophy*, (Oxford University Press, 1992).

Sarojini Henry (2009), *Science Meets Faith*, (St. Pauls Mumbai 2009)

Schilpp, Paul Arthur (Editor) (1941), *The Philosophy of Alfred North Whitehead*, (North Western University, 1941)

Schmidt, Paul F. (1967) *Perception and Cosmology in Whitehead's Philosophy*, (Rutgers University Press, New Jersey, 1967)

Sen, Amartya (1990), *On Ethics and Economics*, (Oxford University Press, Delhi, 1990)

Shaffer, Jerome A. (1994), *Philosophy of Mind*, (Prentice Hall of India, New Delhi, 1994)

Sheldrake, Rupert (2013), *The Science Delusion*, (Coronet, Hodder & Stoughton Ltd, London, 2013)

Sherburne, Donald W. (editor) (1965), *A Key to Whitehead's Process and Reality*, (Indiana University Press, London, 1965)

Smolin, Lee (1998), *The Life of Cosmos*, (Phoenix paperback edition, Great Britain,1998)

Smolin, Lee (2008), *The Trouble with Physics*, (Penguin Books, London, 2008)

Solso, Robert L. (2005) *Cognitive Psychology,* (Pearson education, Delhi, second Indian reprint, 2005)

Sweet, William (2003), *Religion, Science and Non-science,* (Dharmaram Publications, Bangalore, 2003)

Tarnas, Richard (1991), *The Passion of the Western Mind,* (Pimlieo, London, 1991),

Taylor, Richard (1994), *Metaphysics,* (Prentice- Hall of India Pvt.Ltd, New Delhi, 1994)

Teilhard Chardin (1999), *Phenomenon of Man,* (Sussex Academic Press, 1999).

Thilly, Frank (2000), *A History of Philosophy,* (SWP Publishers, New Delhi, 2000).

Thomas.A. P, (General Editor) (2012), *Cell and Molecular Biology – The Fundametals,* (Green Leaf Publications, Kottayam, Kerala, 2012)

Thomson, Mel (1997), *Philosophy of Religion,* (Teach Yourself Books, UK, 1997)

Urmson J.O. and Jonathan Ree (1989), *The Concise Encyclopedia of Western Philosophy and Philosophers,* (Unwin Hyman, London, 1989).

Vijayakumaran Nair and Jayaprakash (2007), *Cell Biology Genetics Molecular Biology,* (Academica, Thiruvananthapuram, Fourth Edition)

Washburn, Phil (1997), *Philosophical Dilemmas: Building a Worldview,* (Oxford University Press, New York, 1997)

Whitehead, Alfred North (1978), *Process and Reality,* (Original Edition 1929; Corrected Edition by David Ray Griffin and Donald Sherburne, New York, The Free Press,1978)

Index of Names

Anaxagoras, 36

Anaximander, 35

Anselm, St. (1033-1109), 41, 91

Anthony Harrison– Barbet, 13, 33

Aquinas, Thomas, St. (1224-1274), 41, 56, 91

Aristotle (BC 384-322), 39, 41, 55-8, 67-70, 91, 94, 97, 119, 125, 151

Augustine, St. (354-430), 41, 56

Ayer, A.J. (1910-1989), 29

Bacon, Francis (1561-1626), 47

Barbour, Ian, 82

Bertalanffy, 108

Berkeley, George (1685-1753), 48, 56, 100-1

Bohm, David (1917-1994), 83, 107

Brooke Noel M, 13, 32, 33, 108

Bruder, Kenneth, 13, 33, 108

Capra, Fritjof, 83

Caputo, John D, 169

Charlesworth, Max, 33, 59, 106

Chardin, Teilhard de, 43, 56, 76-77, 107

Chatterjee, Margaret, 13, 33, 57, 169

Confucius (BC 551-479), 105

Copleston, Frederick S.J., 13, 33

Darwin, Charles (1809-1882), 49, 76, 77, 118

Davies, Brian, 33

Davies, Paul, 112

Democritus, 36, 46

Dennett, Daniel, 169

Derrida, Jacques (1930-2004), 52, 56

Descartes (1596-1650), 47, 56, 71, 73, 81, 90-94, 99, 126, 169

D'Souza, Richard, 107

Einstein, Albert (1879-1955), 50

Empedocles, 36

Ewing A. C. 13, 33, 108, 169

Fichte, Johann Gottlieb (1762-1814), 74

Foucault, Michel (1926-84), 56

Francis Abraham, 147

Frost. S. E , 124, 127

Galileo (1564-1642), 90

George Thomas White Patrick, 13, 32, 33, 125

Gettier, Edmund, 30, 31

Goethe (1749-1832), 74

Grayling A.C., 13, 33, 106, 108

Griffin, David Ray, 77, 107

Guyer, Paul, 33

Hartshorne, Charles (1897-2000), 43, 77, 81

Haught, John F., 77, 107

Hawking, Stephen, 112

Hegel, Georg Wilhelm, 43, 56, 74-79, 88, 125

Heidegger(1889-1976), 56, 89

Heraclitus (BC 535-475), 35, 49, 56, 70

Hick, John H. , 108

Hicks, John, 147

Hiriyanna M, 108

Hospers, John, 13, 169

Hume, David (1711-1760), 48, 56, 101-3, 153, 160

Husserl, (1859-1938), 86-87

Inkeles, Alex, 147

Kant, Immanuel (1724-1804), 33, 47, 56, 66, 74-5, 93-97, 101, 106, 108, 114, 117, 137

Kierkegaard (1813-1855), 88

Kuhn, Thomas (1922-1996), 56

Kunkolienker, Kamladevi, 107

Laszlo, Ervin, 83-84, 107-8

Lavin.T. Z. 13, 33, 106

Leahey, Thomas Hardy, 169

Leibniz, Gottfried Wilhelm (1646-1716), 43, 56, 73-74, 78-81, 125

Leucippus, 36, 46

Levinas (1905-1995), 56

Levin, William C, 133, 147

Locke, John (1632-1704), 48, 56, 99-100

Lucy Mair, 147

Luke, George (1953 -), (see Prologue of this book)

Macquarrie, John, 33, 106, 107

Martin Hollis, 147

Masih .Y, 33

Michael Haralambos, 147

Newton, Isaac (1642-1727), 90, 93, 101

Nietzsche (1844-1900), 56, 88

Oommen. M .A., (see Prologue of this book)

Parmenides (BC 515), 35, 39, 60

Plato (BC 428-348), 36, 39, 41, 55-71, 91-5, 106, 125

Plotinus (AD 205-270), 56, 71

Pulingandla R, 108

Pythagoras (BC 580-500), 35

Radhakrishnan S (1888-1975), 108

Ree, Jonathan, 13, 33

Robert John Russell, 107

Rousseau (1712-1778), 74

Russell, Bertrand (1872-1970), 51

Sartre, Jean Paul (1905-1980), 56, 89

Schelling, Friedrich Wilhelm (1775-1854), 74

Schiller (1759-1805), 74

Sen, Amartya, 147, 170

Socrates (BC 469-399), 36, 39, 56

Spinoza, Benedict (1632–1677), 33, 43, 56, 71-4, 79, 81, 125

Sophists, 36, 70

Tarnas, Richard, 13, 33, 56, 106

Thales (about BC 640-546), 35

Thilly, Frank, 13

Thomson, Mel, 13, 33

Urmson J.O., 13, 33

Virk, Hardev Singh, (see Introduction of this book)

Washburn, Phil, 13, 33

Whitehead, Alfred North (1861-1947), 43, 56, 76-82, 107

William James, 48, 156

Wittgenstein, Ludwig (1889-1951), 51, 56

Zeno (BC 334-262), 56, 70

Index of Subjects

abbreviation, 102, 103, 160

Absolute Mind, 75

Advaita Vedanta, 85

Agnosticism, 96

allegory of the cave, 66

alphabets, 2

analytical geometry, 114

analytical philosophy, 50, 51

artistic life system (ALS), 124, 136, 139

atom, 21, 31, 36-7, 46-50, 73, 83, 104, 115-9, 160, 166

atomism, 36, 46

axiom, 25, 71, 89, 91-2, 121, 126

Ayurveda, 83

belief, 3-5, 29, 30, 37

big bang, 115

black theology, 52

body-mind dualism, 71, 73, 76, 78, 81, 92, 119, 126

BNS, 118-9, 143, 161

Brahman-Maya, 80, 85, 125

Buddhism, 43, 82, 85, 105

cause-effect relation, 8, 10, 16, 25, 50, 70, 92, 95, 101-2, 110, 142, 153-4, 162

Christianity, 41, 69, 139

Christian mysticism, 43

classical science, 6, 46-7, 103-4, 155

classification of knowledge, 2-13, 25, 35, 38, 54, 56, 129, 143, 150

Cartesian Theater, 149, 169

cogito ergo sum, 91, 92, 126

cognitive mind, 9, 17, 32, 40

coherence theory, 151, 153, 162, 167, 169

complex idea / proposition, 8, 10, 99, 101, 122, 152, 157

computer model functionalism, 49, 118

Confucianism, 105

consciousness, 21, 22, 40, 42, 62, 74-89, 98-9, 106, 111-126, 134-6, 143, 161, content view, 6-9, 27, 33-42, 49, 63, 70-4, 81, 86-7, 111-8, 120-126, 145, 151-6, 169

context, 7, 10, 42, 49, 51, 52

contingent truth, 97, 149, 154-168

correspondence theory, 61, 66, 154, 156, 162, 167, 169

cosmology, 50, 115, 155

critical realism, 82

culture, 36, 49, 51-6, 74, 79, 129, 130-142, 156

Darwinism, 118

deconstruction, 52

deduction / deductive, 8, 10, 24-32, 44-7, 53, 64-9, 80-3, 90-103, 114, 122, 142-168

deism, 46-7, 54-8, 77, 89-93, 112-3

Divided Line. 63-5, 106, 124

Dualism, 22, 46, 62, 66, 74, 98

eastern philosophy, 54, 58, 104-8

economic life system, 123, 136

ecnomic philosophy, 20

economics, 12, 114, 122, 139, 143, 147, 166-7, 170

empirical mysticism, 44, 58, 82-5, 89

empirical spiritual process, 44, 54, 84

empiricism, 27, 47-8, 55, 58, 66, 97-104, 134, 145, 153-9, 164, 167

Enlightenment, 48, 99

environmental theology, 52

epiphenomenalism, 55, 97-9, 101, 103

epistemology, 12-16, 20-33, 40, 44, 47, 51, 63, 66, 80-9, 103-9, 157

ethical life system (ETLS), 123, 136, 167

ethics, 19, 36, 59, 71, 88, 122, 124, 139, 143, 147, 150, 165-170

evil, 39, 63, 89, 124, 127, 169

evolution, 7-21, 34, 38, 45, 48-9, 63, 75-7, 112-126, 155

evolutionary theology, 77

existence, definition of, 3-7, 16-23, 28-53, 94

existence is not a predicate, 94, 137

existentialism, 44, 49, 52, 55, 87-89

experimental syllogism, 153

fact, 6

factory, 27, 96, 114, 116, 121-2, 125, 136, 164

faith, 1, 5, 18, 59, 62, 64-5, 76, 79

fallible, 29, 30149

family resemblance, 37, 53, 130, 135

faminist theology /philosophy, 52

forms of life, 51

God, 4, 8-10, 14, 18, 25-46, 59-92, 123-127

good and bad, 15, 39, 122, 128, 134-5, 146, 165-8

habit of association of ideas, 101, 102

Hinduism, 80, 85, 105, 125, 139

Hindu mysticism, 43, 82

historicism, 42, 55, 76, 88

hologram, 83

hypothesis, 10, 24, 53, 80, 92, 114, 149

idealism, 36, 40-3, 47, 54-74, 87-91, 101, 112, 123

Idea of the Good, 62, 64

induction / inductive, (see deduction / deductive)

infallibilism, 30, 31, 149

inference, 10, 12, 21, 24-28, 31, 36, 47, 53, 80-5, 92, 103, 114, 120-1, 137, 145-6, 153-5, 160, 163

Inference to the Best Explanation (IBE), 112, 159

information, 3, 4, 8, 26, 116-8, 151

institution, 76, 128, 131-140, 146

instrumentalism, 162

intellect, 35, 63-66

intellectual mind, 17, 163

intelligent design / designer, 53, 54, 112-3, 116-8, 123, 126

interactionism, 73, 98

intuition, 63-66

isoquant, 122

justification, 2, 12, 24, 28-32, 40-8, 53, 64-9, 80-104, 117-137, 146-167

justified true belief, 29, 30, 148, 157

Kabbalah, 43, 82

kingdom, 10, 12, 23, 136, 137

knowledge, definition of, 2-5

language game, 51

law of complexity consciousness, 76

layered view, 143

liberation theology, 52

life, 1, 6, 9, 18-21, 29, 39, 42, 44, 63, 71, 78, 84, 88-93, 105-124, 128-150, 163-7

Life and Mind, 15, 17, 108, 112, 114, 124, 126-7

life system, 128-147

linguistic analysis, 50, 51, 55

logical positivism, 27, 51, 55, 59, 121

machine-algorithm, 50, 117, 118

materialism, 36, 46-7, 53-8, 65, 73-8, 87, 97-8, 100-5, 112, 115, 123, 134, 145

material process, 42, 48, 70-81, 98

matter and energy, 16-22, 28, 31, 37-50, 61-119

matter-energy system, 115, 116

Maya, 80, 125

mechanistic worldview, 34, 38, 45-48

metaphysical realism, 40, 53, 55, 61, 66, 69-82, 92-4, 106, 123

metaphysics, 19, 32, 37-55, 67-9

methodological stages, 10, 12, 31

methodology, 12, 24-31, 44, 48

mind, 1-9, 17, 21-41

mind-body dualism (see body-mind dualism)

momentariness, 85

monadology , 42, 55, 73

monism, 22, 33, 72

monistic philosophy, 22, 113

monotheism, 41, 62, 69

multiple-realisability, 50

mystical mind, 17, 41, 43, 54, 107, 124, 142-3, 163-4

mystical realism, 32, 84, 89

mystic cults /religions, 43, 44

mysticism, 7, 18, 27, 35, 40-5, 54-8, 66-73, 80-90, 105, 150, 163-4

mystic social knowledge, 142-3, 164-6

mystic worship theology, 44

naïve realism, 99, 104

naturalism, 47, 55, 97, 101, 104, 117, 145

natural life system (NLS), 123, 136-139

necessary-contingent truth, 161-8

necessary truth, 93-4, 149, 151-163, 167-8

neo-Platonism, 42, 71, 125

nervous system, 118, 143, 161

New Age spiritual cults, 52

nirvana, 80, 85

nonbeing, 45, 61, 66-7, 75

normal social science, 139, 142-3, 147

nothingness / void, 80, 85

noumenon, 74, 96

OMEGA, 107

ontology, 13-23, 40, 43, 53, 59-62, 70-77, 87, 95, 105, 109-111, 158, 161

organic levels of knowledge, 12

organic world view, 34-41, 54-59

organism, 10-12, 29, 38-47, 73-79, 92, 114-8, 137-8, 156

Origin of Universe, 15, 83, 102, 115, 169

panentheism, 42, 43, 55, 77

panexperiantialism, 78

panpsychism, 73, 78, 81

pantheism, 41-3, 54-7, 71-88, 105, 07, 127

parallelism, 73, 81

paramporul, 120, 121

parapsychology, 78, 164

particle-wave duality, 50, 83

phenomenalism, 100, 101

phenomenology, 44, 49, 52, 55, 85-7

philosophy, definition of, 11-20

philosophy of mind, 27, 49, 66, 69, 81, 85, 97, 118-9, 128, 143, 146, 159

philosophy of religion, 40, 57, 89, 90, 127

philosophy of science, 59, 96, 115, 150, 169

physicalisation, 115, 145

physical laws, 10, 45-7, 79, 90-6, 103, 155, 162, 168

physical process, 34, 37-8, 42, 46-57, 78, 79, 97-101, 107, 111, 116, 145, 149, 155-6

pluralism, 22, 52, 63, 104, 105, 115, 134

political philosophy, 20, 59, 105, 123, 136, 139, 140

polytheism, 59, 62-5, 105, 106

post modernism, 52, 156

pragmatic theory of truth, 155-6

pragmatism, 48, 118, 138, 161-2

problem of evil, (see evil)

problem of induction, 103, 112, 149, 153-4, 160-1, 168

process philosophy, 42, 44, 58, 70, 72, 74, 78, 81

process theology, 43-4, 70, 76-81, 89, 107

process reality, 43, 44, 82

process view, (see content view)

production function model, 114

proposition, definition of, 3

purpose, 38-48, 59, 62-3, 68-72, 99, 112, 116-7, 122-134, 156, 165-8

puzzle of life, 117

quantum physics, 7, 12, 31, 49-51, 83-4, 107, 155

rational, 7-9, 19, 25-6, 34-46

rationalism, 27, 40, 89-97

rational mysticism, 44, 58, 70, 80-82

realism, 32, 84, 94-7, 106, 112, 114, 119, 120, 164

reality and phenomenon, 16, 40. 60, 63, 74, 109-125

reason, 64-6, 70, 79, 93

relativism, 36, 52, 156

religious life system, 136-140

religious mind, 17, 28, 139

religious truth, 148, 157, 163-9

Renaissance, 45

representative realism, 100

Republic, 59

role / social status, 130

Roman Catholic Church, 41, 76

rules of thought, 69-70, 97, 106, 117, 119, 123

science-religion conflict, 13

scientific mind, 17, 28, 74, 96, 107, 142-5, 160

scientific realism, 32, 55, 59, 104, 112, 115, 159

scientific revolution, 45-6, 90

scientific truth, 148-162

self / non-self (anatman), 74, 80, 85-6, 96, 126, 149

self-consciousness, 74, 86, 99

self-interest and society-interest, 134-140, 165-8

social, 128-148

society, 130, 132

sociology, 138-144

socratic method, 56

soul, 14, 19, 25-8, 31, 37, 55, 59, 65-79

source, 2, 12, 24, 27-8, 65-71

space and time, 96

species, 10-12, 123, 137

Spirit, 72, 74-79

spiritual process worldview, 41-44, 70-80

spiritual science, 11, 12, 83, 164

Stoicism, 42, 55, 70-71

stuff of universe, 20, 35

subject and predicate, 69, 93-97

subjective idealism, 101

substance, 35-79, 94, 100-100

Sufism, 43, 82

superscientific, 17, 123

syllogism, 26, 149, 151-3, 159

system model of knowledge, 114-128, 134, 142-5

system model of truth, 162-168

system Model of Ultimate Reality, 109, 119-125

system Philosophy, 13, 22, 57, 106-9, 113-9, 123-8, 134, 139-150, 157-169

Taoism, 43, 82, 85, 105

taxonomy, 5, 9, 129

teleology, 39, 62-3, 67, 69

theism, 41, 42, 54-5, 62

theology, 9, 12, 18, 19, 41-4, 51-2, 70, 76-84, 89-91, 107, 112, 123

theory of forms, 61-7, 70, 106

theory of four causes, 68, 69

theory of knowledge, 1, 9-33, 58, 80-2, 90, 99

theory of reality, 16, 20, 36, 40, 41, 46, 53, 60-7, 83, 95, 109-111, 119, 148, 162

thesis-antithesis-synthesis, 76

thing-in-itself, 74, 96

three levels of nature, 114, 119, 126

tree of knowledge, 9

truth, 2, 12, 15, 17, 24, 26, 29-32, 44, 48, 61, 66, 74, 80, 91-4, 97, 110, 121-4, 146-170

TyHDTI scheme, 31, 36, 53, 114, 120-1, 143, 145, 149, 156-168

ultimate reality, 17, 57, 84, 109-113, 119-128, 146, 162

underdetermination, 161

universal, 60, 66, 70, 71-3, 83, 87, 92, 132, 133

value, 6-8, 18, 37-45, 62-4, 122, 129-153, 165-8

Vedanta, 80, 85, 105, 125

western philosophy, 18, 34-41, 54-8, 85, 93, 106, 109, 111

worldview, 16, 23, 32-57

yin-yang, 105

www.ingramcontent.com/pod-product-compliance
Lightning Source LLC
Chambersburg PA
CBHW030927180526
45163CB00002B/491